Advances in Electronic
Ceramic Materials

Advances in Electronic Ceramic Materials

A collection of papers presented at the 29th International Conference on Advanced Ceramics and Composites, January 23-28, 2005, Cocoa Beach, Florida

Editors
Sheng Yao
Bruce Tuttle
Clive Randall
Dwight Viehland

General Editors
Dongming Zhu
Waltraud M. Kriven

Published by

The American Ceramic Society
735 Ceramic Place
Suite 100
Westerville, Ohio 43081
www.ceramics.org

Advances in Electronic Ceramic Materials

ISSN 0196-6219

ISBN 1-57498-235-4

Contents

Advanced Dielectric Materials Phenomena

Microwave Dielectric Materials

General Topics in Electronic Ceramics

Preface

The Electronics Division of the American Ceramic Society sponsored two different electronic ceramics symposia at the 29th International Conference on Advanced Ceramics and Composites organized by the Engineering Division of the American Ceramic Society. This proceedings contains selected papers from the two symposia: (1) Emerging Sensor Technologies Based on Electroceramics and (2) Advanced Dielectric, Piezoelectric and Ferroelectric Materials.

The symposium on Emerging Sensor Technologies Based on Electroceramics proved to be a magnificent opportunity to discuss the latest breakthroughs in the field of sensor science and technology. With over 80% of the top scientists in the world in the ceramic sensor field gathered in one room on beautiful Cocoa Beach, it was truly a monumental event. Nowadays, sensor technology is widely used in many applications including critical care, industrial process control, emission monitoring, automotive and home security systems, and more recently, for homeland security. It has undoubtedly made a considerable impact on modern society and human life. The economic and social benefits of sensor technology are evident and multifaceted. In particular, because of the excellent thermal stability of ceramic materials, ceramic sensors have been shown to provide a promising gas detection methodology in harsh environments such as engine and fuel cell exhausts. These currently receive much attention for environmental protection and hence became one of the important topics in this symposium.

Sensor technology is largely dependent on the progress made in the field of material science. In this sensor symposium, for example, several reports discussed new catalysts to modify sensing electrodes of non-Nernstian type sensors and semiconducting ceramic sensors; big improvements in sensing performance have been achieved. Several new solid electrolyte materials have been developed and showed to be promising for constructing sensors. New nano-structured materials and SiC based high temperature ceramics showed unique properties for chemical sensing. Therefore, great progress in sensor technology has been made through developing and employing novel materials. In fact, sensors are becoming increasingly important in ceramic material applications. Constructive discussions and communications such as this symposium will advance unequivocally our understanding and perspective on sensor science.

The Advanced Dielectric, Piezoelectric and Ferroelectric Materials Symposium brought together researchers and engineers from Europe, Asia, North America and Australia to present and exchange ideas on the latest scientific and technical developments in the field of dielectric, ferroelectric and piezoelectric ceramics. Seventy-four papers and posters were presented. A particular strength of this meeting was the ample opportunity for informal, lively discussions of technical subjects among the meeting participants. Among the topics that attracted a great deal of interest were the latest materials science discoveries for high dielectric constant copper calcium titanate materials, novel high temperature piezoelectrics, fundamental understanding of crystal structure and sub-nanometer distortions on microwave properties, and integration of Cu electrodes with BaTiO3 and PZT based thin films.

We gratefully acknowledged the efforts of the organizing committee for putting together a technical program that drew leading technical experts throughout the world as symposium speakers. We are also extremely grateful to all the speakers who contributed to this highly successful symposium, especially those from outside of North America.

Sheng Yao
Emerging Sensor Technologies Based on Electroceramics

Bruce Tuttle, Clive Randall and Dwight Viehland
Advanced Dielectric, Piezoelectric and Ferroelectric Materials

Emerging Sensor Technology
Based on Electroceramics

ZIRCONIA-BASED GAS SENSORS USING OXIDE SENSING ELECTRODE FOR MONITORING NOx IN CAR EXHAUST

N. Miura, J. Wang, M. Nakatou, P. Elumalai, S. Zhuiykov and D. Terada
Art, Science and Technology Center for Cooperative Research, Kyushu University,
Kasuga-shi, Fukuoka 816-8580, JAPAN
T. Ono
R&D Division, Riken Corporation, Kumagaya-shi, Saitama 360-8522, JAPAN

ABSTRACT

Solid-state electrochemical sensors using yttria-stabilized zirconia (YSZ) and oxide sensing electrode (SE) were fabricated and examined for NOx detection at high temperatures. Among various single-oxide SEs examined, NiO-SE for the mixed-potential-type NOx sensor was found to exhibit rather high sensitivity to NO_2 even in the high temperature range of 800-900°C. This sensor showed quicker response and recovery transients in the presence of water vapor, compared with that in the dry sample gas. The sensing mechanism of this type of sensor was discussed on the basis of the catalytic activities to the electrochemical and non-electrochemical reactions. It was also shown that the new-type complex-impedance-based (impedancemetric) NOx sensor attached with $ZnCr_2O_4$-SE exhibit good sensing characteristics to NOx at 700°C. Furthermore, the sensitivity to NO was almost equal to that to NO_2 in the concentration range from 0 to ca. 200 ppm at 700°C. A linear dependence was observed between the sensitivity of the impedancemetric sensor and the concentration of NOx even in the presence of 8 vol. % H_2O and 15 vol.% CO_2. The planar laminated-type structure for the impedancemeteric NOx sensor was proposed for protecting NOx sensitivity from the influences of the co-existing combustible gases as well as the change in oxygen concentration in exhaust gas.

INTRODUCTION

Recently, the demand for reliable, compact, low-cost solid-state sensors, which are capable of detecting nitrogen oxides (NOx) in different application, has been enhanced substantially. This demand has been driven by strong recent legislation in EU, USA and Japan. For example, in UK under Air Quality Regulations (1997) for NOx, standards of 150 *ppb* (hourly maximum) and 21 ppb (annual average) must be achieved by the end of 2005.[1] On the other hand, according to a new report by Fredonia Group, US demand for sensor products (including sensors, transducers and associated housing) is projected to increase 7.8% annually to \$13.8 billion in 2008.[2] For monitoring in the automotive exhausts, the sensor should be able to detect NO_x concentration from 10 ppm up to 2000 ppm in very harsh environment where the temperature constantly fluctuates from 600°C up to 900°C, since the temperature of engine may occasionally reach up to 900°C, during vehicle acceleration. It is therefore vital to investigate thoroughly the SE materials of the NOx sensor in order to provide high NOx sensitivity and selectivity, long-term stability at high temperatures as well as fast response and recovery for a practical sensor.

Last decade, an ultra lean-burn (or direct-injection type) engine system has been developed to improve fuel efficiency as well as to reduce CO_2 and NOx emissions from engine. In this engine system, newly-developed NOx-storage catalyst should be used for compensating the low NOx-removal ability of the conventional three-way catalyst under the lean-burn (air rich) condition, as shown in Fig. 1.[3] It is important, therefore, to have high-performance NOx sensors installed

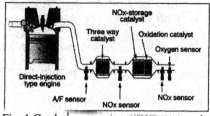

Fig. 1 Catalytic converter system equipped with NOx sensors for the exhaust gas emitted from a new-type car engine.

at the point after (or both before and after) the NOx-storage catalyst for such a system. The mixed-potential-type NOx sensors based on YSZ and oxide SE have been receiving considerable attention among the others YSZ-based NOx sensors,[4] as potential candidates for practical sensor measuring car emissions. For example, last few years the most of the research groups have focused on the development of oxide SEs which are capable of working at high temperatures in car exhaust.[5-9] The use of oxide SE in this type of NOx sensor was found to be very effective for sensitive and selective NOx measurement at high temperature.[10] However, there are only a few designs of the NOx sensors reported to date[11-16] that can monitor total NOx (NO + NO$_2$) at high temperatures regardless of NO$_2$/NO ratio in real exhausts. One of these sensors is the complex-impedance-based NOx sensor originally designed and developed by our group.[15, 16] In addition, the most of results for oxide SEs published so far have revealed that the NOx sensors using the oxide SEs show relatively good sensing characteristics only in the temperature range of 450-700°C. The majority of these sensors have difficulty to operate at temperatures higher than 700°C. Such a higher limitation of operating temperature is caused by the substantial decrease in the NOx sensitivity with increasing temperature. Based on the above-mentioned facts and keeping it in mind that there are no commercial high-temperature NOx sensors available on the market at the moment, further search for oxide SE has been done. As a result, it was found quite recently that NiO is a very effective SE for measuring NOx concentrations at temperatures higher than 800°C.[17, 18] There were only limited number of publications reporting the properties and sensing characteristics of oxide SE that can measure NOx concentration at temperatures over 800°C.[17, 18] Furthermore, we have also discovered that the new complex-impedance-type NOx sensor using ZnCr$_2$O$_4$-SE shows rather good sensing characteristics for measurement of total NOx concentration at temperature as high as 700°C. Thus, we report here the sensing properties and sensing mechanism of the mixed-potential-type YSZ-based NOx sensor attached with NiO-SE at temperatures of 800-900°C as well as the main sensing performances of the new complex-impedance-type NOx sensor using ZnCr$_2$O$_4$-SE.

EXPERIMENTAL

A tubular NOx sensor was fabricated by using a commercial one-closed-end YSZ tube (8 wt.% Y$_2$O$_3$-doped, NKT). The tube is 300 mm in length and 5 and 8 mm in inner and outer diameters, respectively. NiO-powder paste was applied on the outer surface of the YSZ tube and sintered at 1400°C to form a sensing electrode (SE). A Pt lead wire was wound around the oxide layer to make a good electrical contact. Pt paste was painted onto the inside of the YSZ

4

tube and fired afterwards at 1000°C for 2 h to form a reference electrode (RE). Figure 2 shows the cross-sectional (schematic) view of the obtained tubular YSZ sensor attached with NiO-SE and Pt-RE.

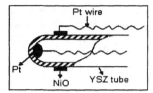

Fig. 2 Cross-sectional (schematic) view of the tubular YSZ sensor.

The fabrication of planar NOx sensor was done by the use of YSZ plates (8 wt. % Y_2O_3-doped, 10 x 10 mm; 0.2 mm thickness). Pt paste was printed on both sides of the YSZ plate and then was fired at 1000°C for 2 h in air. On the one side of the YSZ plate, two rectangular Pt-stripes were formed as Pt-RE of the sensor and, on the other side; six narrow Pt-stripes were formed as a base (current collector) for the NiO-SE film. NiO powder (Kojundo Chemical Lab. Co. Ltd., 99.97% purity) was thoroughly mixed with α-terpineol (20 wt.%) and the resulting paste was applied on the top side of the YSZ plate attached with narrow Pt-stripes by means of screen-printing technique to form SE. The planar sensors were sintered at 1100°C, 1200°C, 1300°C and 1400°C, respectively, for 2 h in air. Although the Pt-RE of the tubular YSZ-based NOx sensors has been always exposed to the atmospheric air during experiments,[19] the Pt-RE of the present planar sensor is always exposed to the measuring gas. Pt wires were spot-welded onto the Pt connecting-spots of both SE and one of REs to provide the good electrical contact with a measuring equipment.

The microstructure and surface topography of the NiO-SE films were examined by using an SEM (JEOL electron microscope, JSM-340F) operating at 15 kV. The crystal structure of the films was studied by means of a wide-angle XRD (RIGAKU X-ray diffractometer, RINT 2100VLR/PC). The CuKα radiation (λ=1.5406) and 0.5°/min angle step were used for the XRD measurement. The YSZ/NiO interface was observed by the use of a TEM (FEI Inc., Model TECNAI F20) at the Research Laboratory for High Voltage Electron Microscopy of Kyushu University. The accelerated voltage was 200 kV for all experiments. BET surface area was measured by using an automated gas-sorption system (Quantachrome Autosorb, version 1.20).

NOx-sensing experiments were carried out in a conventional gas-flow apparatus operating in the temperature range of 700-900°C, as shown in Fig. 3. The sample gas containing various concentrations of NO (or NO_2) was prepared by diluting with dry nitrogen and oxygen gases and was allowed to flow over the sensor at a constant flow rate of 100 cm^3 min^{-1}. The concentrations of NO and NO_2 were changed from 10 to 400 ppm. The difference in potential (emf) between NiO-SE and Pt-RE of the sensor was measured by a digital electrometer (Advantest, R8240). The base gas was 5 or 20.9 vol.% of dry O_2 diluted with dry N_2, and the sample gas was 10-400 ppm of NOx diluted with the base gas. In

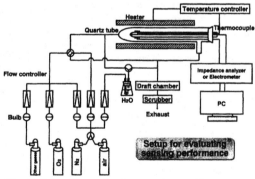

Fig. 3 Setup for evaluating sensing performances of NOx sensors.

order to humidify the base gas and the sample gas, 5 vol.% water vapor was mixed with them. For the impedance-based sensor, the complex impedance and the phase angle between SE and RE were measured with a complex impedance analyzer (Solarton, 1255 WB) in the frequency range of 0.01 Hz - 1 MHz to obtain complex-impedance (Nyquist) plots. As an output signal, the complex impedance value ($|Z|$) was used and was monitored at a fixed frequency of 1 Hz.

RESULTS AND DISCUSSION

Crystal Structure and Morphology of the Oxide SEs

One of the most important factors that affect the sensing characteristics of the mixed-potential-type NOx sensors is the composition or morphology of SE. XRD patterns for both NiO and $ZnCr_2O_4$ thick films sintered at 1400°C and 1200°C, respectively, contained only peaks of crystalline face-centered cubic NiO (JCPDS PDF#47-1049) phase or $ZnCr_2O_4$ (JCPDS PDF#22-1107) phase. It is seen that all peaks for the sintered NiO and $ZnCr_2O_4$ were narrow. This suggests the excellent thermal stability of NiO and $ZnCr_2O_4$.

Figure 4 shows SEM micrographs of the cross-section of the screen-printed NiO-SE films sintered at different temperatures on the YSZ substrate. The thickness of films was maintained at almost same value of about 7 µm even after sintering at different temperatures. With decreasing sintering temperature, the average pore size was found to increase from ca. 0.5 µm (for the 1100°C-sintered film) to ca. 2 µm (for the 1400°C-sintered film). This suggests that higher sintering temperature gives lower number of reaction site at a triple-phase-boundary (TPB) of gas/YSZ/NiO. These observations also show that the high sintering temperature provides relatively low surface-to-volume ratio on SE layer which consequently brings low catalytic activity to electrochemical reactions. The less surface area of the NiO film will also lead to lower heterogeneous catalysis (decomposition of NO_2 to NO). In the case of the $ZnCr_2O_4$–SE sintered at 1200°C, it was also relatively porous with an average grains size of about 0.2-0.8 µm.

TEM image of the TPB (gas/YSZ/SE) for the NiO-SE sintered at 1400°C onto YSZ substrate is presented in Fig. 5. Analysis of this figure showed that the TPB, where all electrochemical reactions occur, is a small

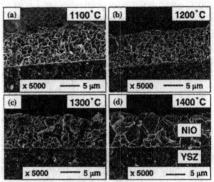

Fig. 4 SEM micrographs of the cross-section of the screen-printed NiO films (on YSZ substrate) sintered at various temperatures.

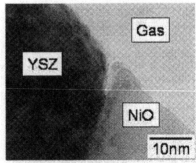

Fig. 5 TEM micrograph of the TPB at gas/YSZ/NiO for the NiO film sintered onto YSZ substrate at 1400°C.

6

curve which goes along with the YSZ/NiO interface. This nano-scale picture clearly shows that the YSZ/NiO interface is about 2 nm in thickness and goes along the boundary between the NiO and YSZ grains joined together. Thus, the TPB consists of the gas and the combination of the relatively small nano-scale curves along the YSZ/NiO interface as well as small islands when the YSZ and NiO grains connected to each other by one or a few single points.

Sensing Performances of the Mixed-potential-type NOx Sensor Using NiO-SE

Among the thirteen kinds of single-oxide SEs tested, the NiO-SE was found to give the highest NO_2 sensitivity at 850°C. The YSZ-based NOx sensor attached with NiO-SE showed a strong linear correlation between the gas sensitivity and the logarithm of NO_2 concentration from 50 ppm up to 450 ppm at temperatures of 800-900°C. Figure 6 shows the variation of the output emf values for the present sensor to 400 ppm NOx (NO or NO_2) when the operating temperature was changed from 800°C up to 900°C. The oxygen concentration was fixed at 20.9 vol.%. It is clear from this figure that the output emf of the NOx

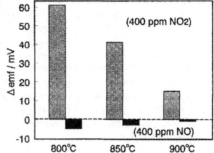

Fig. 6 Variation of the output emf for the NOx sensor using NiO-SE at operating temperatures of 800-900°C.

sensor decreases with increasing operating temperature. However, even at 900°C the sensor attached with NiO-SE gave the emf response of about 15 mV. Such a result is hard to see in the case of any other oxide-SEs tested here and reported to date. Figure 7 shows the response transients to 400 ppm NO_2 under dry and wet conditions for the tubular NOx sensor attached with NiO-SE at 850°C. The emf value was almost zero when the base gas was introduced to the sensor and changed quickly from the base level to the some emf value upon exposure to the sample gas containing NO_2. The 90% response time was about 40 s, however the recovery rate was very slow compared with the response rate. The emf value did not return to the base level and reached only the about 80% recovery level even after 20 min. Such a slow recovery can be explained by the fact that the catalytic activity for electrochemical reaction of oxygen ($\frac{1}{2}O_2 + 2e^- = O^{2-}$) is low for the 1400°C-sintered NiO-SE. Such a slow recovery of the present sensor was found to be improved by humidifying the sample gas. In this test, 5 vol.% of water vapor was incorporated into the dry sample gas. As shown in Fig. 7 (b), the recovery rate was greatly improved after the introduction of water vapor. Under the wet condition, the 90% response and 90% recovery times were about 20 s and about 3 min, respectively. The emf value returned completely to the original level within about 5 min. In addition, the sensitivity (75 mV to 400 ppm NO_2) under

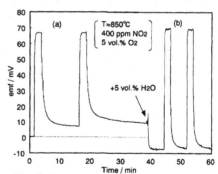

Fig. 7 Response/recovery transients to 400 ppm NO_2 in 5 vol.% O_2 (+N_2 balance) gas in the absence (a) and in the presence (b) of 5 vol.% water vapor at 850°C.

7

the wet condition was a bit higher than that (60 mV) under the dry condition. Moreover, the reproducibility of emf response to 400 ppm NO_2 was quite satisfactory. Thus, the presence of water vapor in car exhausts will give a positive effect to the performance of the present NOx sensor.

Figure 8 shows the comparison of the response transients to 200 ppm NO_2 at 800°C for the planar sensors attached with each of NiO-SEs sintered at various temperatures. It is seen that the steady-state emf value to 200 ppm NO_2 increased from 3 mV to 55 mV when the sintering temperature of SE was increased from 1100 to 1400°C. However, the response rate was lowered by increasing the sintering temperature. These results given in Fig. 8 clearly indicate that the NO_2 sensitivity of the present sensor can be enhanced by increasing the sintering temperature of SE.

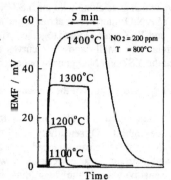

Fig. 8 Response transients to NO_2 at 800°C for the planar sensors using each of NiO-SEs sintered at different temperatures.

Elucidation of Sensing Mechanism for the Mixed-potential-type NO_2 Sensor

The NOx sensing mechanism for the YSZ-based sensors using metal-oxide SE is based on mixed-potential, as has been reported before.[8,17] For the NO_2 sensing, the following electrochemical reactions proceed simultaneously at the interface of YSZ/SE and consequently mixed potential appears on SE:

$$(\text{anodic}) \quad O^{2-} \longrightarrow \tfrac{1}{2} O_2 + 2e^- \qquad (1)$$

$$(\text{cathodic}) \quad NO_2 + 2e^- \longrightarrow NO + O^{2-} \qquad (2)$$

Based on the previously published results,[3,4,10] these reactions occur at different kinetic rates on the dissimilar electrodes. As a result, the emf response of the sensor is the difference in mixed potential on each electrode. Therefore, the catalytic activity to anodic reaction of O_2 (1) should be low and the catalytic activity to cathodic reaction of NO_2 (2) should be high. In order to verify this assumption, the complex impedance measurements were performed at the frequency range of 0.01 Hz ¡ «d MHz in the base gas (5 vol.% O_2 + N_2 balance) at 800°C for the planar sensors attached with each of the NiO-SEs sintered at different temperatures. It was seen that the resistance of electrode reaction in the base gas increased with increasing sintering temperature of SE. This implies that the higher sintering temperature of SE gives lower catalytic activity to anodic reaction of oxygen (1). The decrease in the catalytic activity to the anodic reaction in the case of higher sintering temperature of SE was also confirmed from the results of the polarization-curve measurements. At the same time, the low catalytic activity to the cathodic reaction of NO_2 (2) can be explained using the schematic view of the interface shown in Fig. 9. One can see that smaller NiO grains are present at TPB in the case of lower sintering temperature of SE. In contrast, larger grains are present at TPB in the case of higher sintering temperature. The larger grains may produce less number of reaction sites at TPB. Thus, we can speculate that, in the case of higher sintering temperature, the catalytic activity to

8

the cathodic reaction (2) is low compared to the case of lower sintering temperature, as observed for the anodic reaction.

Moreover, the oxide matrix also plays an important role in deciding the NO_2 sensitivity[17] when we consider the gas-phase reaction:

$$NO_2 \longrightarrow NO + \tfrac{1}{2}O_2 \qquad (3)$$

Based on our previous results,[3,5,19] low conversion of NO_2 in the gas-phase reaction would lead to high NO_2 sensitivity. In the present study, it was observed from the SEM images (see Fig. 4) that the large pores exist in the case of the 1400°C-sintered NiO-SE. As shown in Fig. 9, NO_2 gas makes less contacts with the surface of NiO grains when it diffuses through the large pores presenting in the 1400°C-sintered SE matrix, where its surface acts as a catalyst for gas-phase decomposition reaction of NO_2 (3). Thus, NO_2 can reach the YSZ/SE

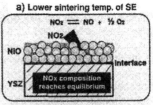

a) Lower sintering temp. of SE

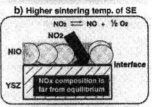

b) Higher sintering temp. of SE

Fig. 9 Schematic views of the effect of both grain size and pore size of the SE matrix on catalytic activities to reactions (1) and (3).

interface without serious decomposition to NO. In contrast, NO_2 makes a significant contacts with the surface of NiO grains when it diffuses through the small pores presenting in the 1100°C-sintered SE matrix where almost all NO_2 gas can be converted into NO before reaching the TPB. Thus, the low catalytic activity to anodic reaction of oxygen as well as less possible conversion of NO_2 to NO in the gas-phase reaction would lead to higher NO_2 sensitivity in the case of the 1400°C-sintered SE. Furthermore, the high catalytic activity for the anodic reaction of oxygen and the high conversion of NO_2 to NO lead to lower NO_2 sensitivity in the case of the 1100°C-sintered SE, in spite of the fact that the high catalytic activity to anodic reaction of oxygen can give faster recovery rate.

All the above results presented here show that NiO-SE gives good sensing performances in the humid exhaust environment even at high temperatures of 800-900°C. Thus, we may conclude that this material is one of the potential candidates for SE of the mixed-potential-type NOx sensor which is capable of detecting NO_x on-board in car exhausts at high temperature.

Sensing Performances of the Complex Impedance-based NOx Sensor

In addition to the mixed-potential-type NOx sensors, our attention during last few years has been focusing on the development of principally new-type YSZ-based sensor for detecting NOx at high temperature.[3,16] In this type of NOx sensors, the change in the complex impedance of the device attached with oxide SE was measured as a sensing signal. Initially we investigated the complex-impedance plots for the devices using spinel-type oxides, such as $CrMn_2O_4$, $NiCr_2O_4$, $NiFe_2O_4$ and $ZnCr_2O_4$, as an SE in base air at 700°C. Both NO_2 and NO (200 ppm each, diluted with dry air) were used as the sample gas during experiments. The sensors attached with the first three oxides were found to provide large and flat semicircular arc in each Nyquist plots in the examined frequency range. The impedance values of these devices were not affected by the existence of NOx under the present condition. However, we found that in the case of the device attached with $ZnCr_2O_4$-SE, the impedance behavior was entirely different from the above-mentioned results. Figure 10 shows how the resistance value (Z', the intercept) at the intersection of the large semi-arc with the real axis at low frequencies (around 0.1 Hz) varies with concentrations of both NO and NO_2. It is seen that the resistance value decreases with an increase in the concentration of both NO and NO_2. Such a behavior is completely different from that for the mixed-potential-type NOx sensor whose response direction to NO is opposite to that to NO_2 (see Fig. 6).

Meanwhile, the Z' value (the intercept, about 2000 Ohm) at the intersection of the large semi-arc at high frequencies (around 50 kHz) did not change even if the concentration of NOx was changed from 10 to 400 ppm. The difference between the impedance ($|Z|_{air}$) in the base air and the impedance ($|Z|_{gas}$)

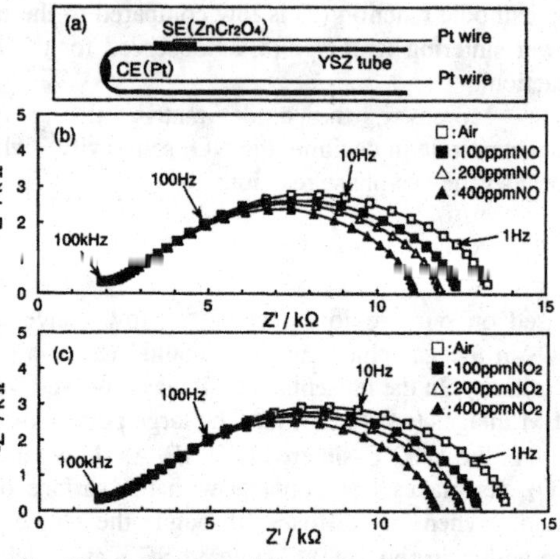

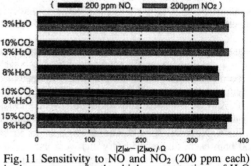

Fig. 10 (a) Cross-sectional view of the complex impedance-based NOx sensor using YSZ and $ZnCr_2O_4$-SE. (b), (c) Nyquist plots in the base air and the sample gas with each of various concentrations of NO and NO_2 at 700°C.

in the sample gas containing NOx at the fixed frequency of 1 Hz has been defined as 'gas sensitivity' of the device.[3, 15,16] We, therefore, investigated the sensitivity of sensor attached with $ZnCr_2O_4$-SE to both NO and NO_2 at high temperature and in the presence of high concentration of H_2O (3 to 8 vol.%) and CO_2 (10 to 15 vol.%), which are usually exist in car exhausts. All measurements were carried out at fixed temperature of 700°C for various NOx concentrations. This investigation revealed the existence of strong linear correlation between 'gas sensitivity' and the measuring NOx concentration from 0 to 200 ppm. Figure 11 shows that the sensitivity of the sensor at 1 Hz to both NO and NO_2 (200 ppm each) was almost constant in the presence of high H_2O and CO_2 concentrations. Moreover, the most interesting result taken into consideration from these tests is the fact that the sensitivity to NO is almost equal to that to NO_2 at 700°C. From the practical point of view, this means that the present device is capable of measuring total NOx (NO and NO_2) concentration in the gas mixture regardless of the NO/NO_2 ratio. This is very valuable point for the development of practical total NOx sensor for car exhausts.

Based on our previous experiences with the mixed-potential-type sensors, we can presume that it

Fig. 11 Sensitivity to NO and NO_2 (200 ppm each) in the presence of rather high concentrations of H_2O (3-8 vol.%) and CO_2 (10-15 vol.%) at 700°C for the YSZ-based sensor using $ZnCr_2O_4$-SE.

is quite possible that the response of the present device is also intervened by the change in O_2 concentration in the sample gas.

In order to verify this assumption, O_2 concentration in the sample gas was changed from 5 vol.% to 80 vol.% at 700°C whilst 'gas sensitivity' of the device to 100 ppm of both NO and NO_2 was recorded. The results of these tests revealed that the value $|Z|$ indeed varied linearly with the logarithm of O_2 concentration at 1 Hz. In the meantime, the 'gas sensitivities' to NO and to NO_2 were almost equal at any O_2 concentration examined. This suggests that the O_2 concentration in the sample gas existing at the space near the oxide-SE of the device should be controlled and should be kept constant during operation. For this purpose, both an O_2 sensor and an O_2-pump could be employed for monitoring and controlling O_2 concentration, respectively. These devices can be built into a new laminated-type YSZ-based device consisting of oxidation catalyst and NO_2 sensing electrode, as shown in Fig. 12. Combustible gases, therefore, can be oxidized by oxidation catalyst in this design of the sensor and the O_2 concentration can be kept constant. Consequently, the NOx sensitivity of the sensor will have no influence by the co-existence of combustible gases and by the O_2 concentration variation in the exhaust gas.

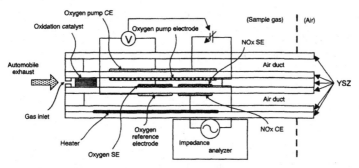

Fig. 12 A cross-sectional view of the proposed laminated-type complex impedance-based NOx sensor.

CONCLUSIONS

First, the tubular and planar YSZ-based sensors attached with each of the NiO-SEs sintered at various temperatures were fabricated for NOx detection at the different environments aiming for monitoring car exhausts. The sensing characteristics of these sensors were examined in the temperature range of 800-900°C. It was found that the NO_2 sensitivity of NiO-SEs was greatly influenced by changing the sintering temperature of SE. Rather high sensitivity to NO_2 was obtained even at 900°C for the sensor using the NiO-SE sintered at 1400°C. The NO_2 sensitivity observed at such a high temperature has never been reported before. The low catalytic activity to anodic reaction of oxygen (1) as well as the scanty conversion of NO_2 to NO on the gas-phase reaction (3) may lead to the high NO_2 sensitivity in the case of 1400°C-sintered SE having larger pores. In opposite, the high catalytic activity to anodic reaction and the high conversion of NO_2 to NO may lead to lower sensitivity in the case of 1100°C-sintered SE having smaller pores and smaller grains. The present investigation indicates that NiO is a promising candidate for the practical SE of on-board planar NOx sensors, although detailed further investigation of the sensing performance for this material is necessary to make a conclusion about its acceptance for the practical use in NOx sensors.

Secondly, the NOx sensing performances of the complex-impedance-based YSZ sensor attached with $ZnCr_2O_4$-SE were investigated at 700°C. The sensing characteristics of the sensor showed rather good sensitivity to NOx concentration from 10 to 400 ppm. The sensitivity to NO was almost equal to that to NO_2 for the present sensor. This means that the present sensor can measure the total NOx concentration in the sample gas. The observed NOx sensitivity was found to vary almost linearly with NOx concentration even in the presence of 8 vol.% H_2O and 15 vol.% CO_2. In addition, a new laminated-type structure for the complex impedance-based NOx sensor has been designed, in order to avoid the effect of the co-existing combustible gases as well as the change in O_2 concentration in exhaust gas.

ACKNOWLEDGEMENT

A part of the present study was financially supported by the Ministry of Education, Culture, Sports, Science and Technology of Japan.

REFERENCES

[1] J. Dixon, D.R. Middleton and R.G. Derwent, "Sensitivity of nitrogen dioxide concentrations to oxides of nitrogen controls in United Kingdom", *Atmospheric Environment*, **35**, 3715-3728 (2001).

[2] Fredonia Group, Inc., "Development of the US Sensors Market", *Final Report*, 1-22 (2004).

[3] N. Miura, M. Nakatou and S. Zhuiykov, "Development of NOx sensing devices based on YSZ and oxide electrode aiming for monitoring car exhausts", *Ceramics International*, **30**, 1135-1139 (2004).

[4] N. Docquier and S. Candel, "Combustion control and sensors: a review", *Progress in Energy & Combustion Science*, **28**, 107-150 (2002).

[5] S. Zhuiykov, T. Ono, N. Yamazoe and N. Miura, "High-temperature NOx sensors using zirconia solid electrolyte and zinc-family oxide sensing electrode", *Solid State Ionics*, **152-153**, 801-807 (2002).

[6] N. F. Szabo, H. Du, S. A. Akbar, A. Soliman and P. K. Dutta, "Microporous zeolite modified yttria stabilized zirconia (YSZ) sensors for nitric oxide (NO) determination in harsh environments", *Sensors and Actuators B*, **82**, 142-149 (2002).

[7] E. Di Bartolomeo, N. Kaabbuathong, A. D'Epifanio, M. L. Grilli, E. Traversa, H. Aono and Y. Sadaoka, "Nano-structured perovskite oxide electrodes for planar electrochemical sensors using tape casted YSZ layers", *J. European Ceramic Soc.*, **24**, 1187-1190 (2004).

[8] F. H. Garzon, R. Mukundan and E. L. Brosha, "Solid-state mixed potential gas sensors: theory, experiments and challenges", *Solid State Ionics*, **136-137**, 633-638 (2000).

[9] E. Magori, G. Reinhardt, M. Fleischer, R. Mayer and H. Meixner, "Thick film device for detection of NO and oxygen in exhaust gases", *Sensors and Actuators B*, **95**, 162-169 (2003).

[10] N. Miura, G. Lu and N. Yamazoe, H. Kurosawa and M. Hasei, "Mixed potential type NOx sensor based on stabilized airconia and oxide electrode", *J. Elecrochem. Soc*, **143**, L33-L35 (1996).

[11]N. Kato, H. Kurachi and Y. Hamada, 'Thick film ZrO_2 NOx sensor for the measurement of low NOx concentration', *SAE Technical Paper Series*, 1998– 980170, 69-77 (1998).

[12]N. F. Szabo and P. K. Dutta, "Strategies for total NOx measurement with minimum CO interference utilizing a microporous zeolitic catalytic filter", *Sensors and Actuators B*, **88**, 168-177 (2003).

[13]M. Hasei, T. Ono, Y. Gao, Y. Yan and A. Kunimoto, 'Sensing performance for low NOx in exhausts with NOx sensor based on mixed potential', *SAE Technical Paper Series*, 2000-01-1203, 1-7 (2000).

[14]T. Ono, M. Hasei, A. Kunimoto, T. Yamamoto and A. Noda, 'Performance of the NOx sensor based on mixed potential for automobiles in exhaust gases', *JSAE Review*, **22**, 49-55 (2001).

[15]N. Miura, M. Nakatou and S. Zhuiykov, "Impedance-based total-NOx sensor using stabilized zirconia and $ZnCr_2O_4$ sensing electrode operating at high temperature", *Electrochemistry Communications*, **4**, 284-287 (2002).

[16]N. Miura, M. Nakatou and S. Zhuiykov, "Impedancementric gas sensor based on zirconia solid electrolyte and oxide sensing electrode for detecting total NOx at high temperature", *Sensors and Actuators B*, **93**, 221-228 (2003).

[17]P. Elumalai and N. Miura, "Influence of annealing temperature of NiO sensing-electrode on sensing characteristics of YSZ-based mixed-potential-type NOx sensor", *Int. Symposium on Chemical Sensors VI: Chemical & Biological Sensors and Analytical Methods*, **8**, 80-88 (2004).

[18]N. Miura, J. Wang, M. Nakatou, P. Elumalai and M. Hasei, "NOx sensing characteristics of mixed-potential-type zirconia sensor using NiO sensing electrode at high temperatures of 800-900°C", *Electrochem. Solid-State Lett.*, **8**, H9-H11 (2005).

[19]N. Miura, K. Akisada, J. Wang, S. Zhuiykov and T. Ono, "Mixed-potential-type NOx sensor based on YSZ and zinc oxide sensing electrode", *Ionics*, **10**, 1-9 (2004).

INTERFACIAL PROCESSES OF ION CONDUCTING CERAMIC MATERIALS FOR ADVANCED CHEMICAL SENSORS

Werner Weppner
Chair for Sensors and Solid State Ionics,
Faculty of Engineering,
Christian-Albrechts University,
Kaiserstr. 2, 24143 Kiel, Germany
e-mail: ww@tf.uni-kiel.de

ABSTRACT
    Ceramic electrochemical gas and fluid sensors are mainly based on potentiometric or amperometric principles. Steady state equilibria are established between the gas or fluid and the surface of the electrolyte or electrode and across a diffusion barrier, respectively. The electrostatic potential drop of the galvanic cells depends on the charging of double layers between the phases in contact with each other. A transfer of electrons and ions has to occur across these interfaces. The kinetics of the various limiting factors, i.e. polarizations, is taken into consideration for determining the partial pressures of single and multiple gas compositions. Time dependent currents and voltages are applied which provide driving forces for diffusion, absorption and charge transfer processes of the electroactive species. These devices are called theta (θ) sensors. Favorably, periodic electrical signals are being employed. The signal - response behavior may be optimized by selecting appropriate frequencies, amplitudes and shapes of the electrical signal as well as favorable chemistry and morphology of the employed materials. Experimental verifications of interface phenomena based ceramic chemical sensors are being presented.

1. INTRODUCTION

    Chemical gas sensors based on ionically conducting ceramic materials have been technologically and commercially extraordinarily successful for many practical applications [1]. The employed class of ceramic materials is mechanically and chemically very stable and may be operated over a wide range of temperatures. Electrical quantities, i.e. voltages and currents, are generated which may be easily measured with high precision and further processed. Especially zirconia-based oxide ion conductors are widely used in automobiles for reduced pollution and lower gasoline consumption, for precisely controlled reduction of ores with improved quality of metals, for hardening steel by carbon via C-CO equilibria and for the control of inert atmospheres in many oxygen sensitive processes.
    The last 40 years have resulted in large progress in our knowledge in the phenomenon of fast ion transport in the solid crystalline and amorphous state and this has resulted in a large number of new ceramic compounds with practically applicable high ionic conductivities which are comparable with common liquid electrolytes. These materials alone, however, are practically

meaningless; only combinations with (two) other materials are important. Ionic junctions with drops of the electrostatic potential are formed at the interfaces with metallic or mixed conducting electrodes, while the bulk materials remain free of electrical fields under open circuit conditions and show only small electrostatic potential drops under current [2]. The electrical fields at the interfaces depend on the chemical potential differences of both mobile species, i.e., ions and electrons. This situation is similar to the case of semiconductors in which the chemical potential difference of the electrons – however, only this one type of charge carriers - between the adjacent phases with different electrical properties plays the key role. Since both ions and electrons equilibrate at the interfaces in the case of ionic devices, the technological difficulties are much larger for the application of solid ionic conducting ceramics for all applications, including fuel cells, batteries, electrochromic systems, thermoelectric converters, photogalvanic cells and also sensors, compared to semiconductors. The mobility of the ions results in variations of the chemical composition of the compounds and even in the formation of new interfacial materials, which generate changes in the performance and causes short life times. Furthermore, the electrical field at the contact between a metallic and a fast ionic conductor is much stronger and extends over a much smaller regime than in the case of a semiconductor junction. This is due to the much higher concentration of charge carriers in metals and ionic conductors compared to semiconductors. The space charge layer is of a thickness in the range of nm instead of µm, which therefore results in a much higher sensitivity with regard to the formation of interfacial layers.

The present work considers the interfacial processes and takes advantage of these phenomena for chemical sensing by ceramic ionic conductors which will lead to advanced devices with the possibility of simultaneous detection of several species simultaneously.

2. FUNDAMENTAL ASPECTS OF INTERFACIAL PHENOMENA

The common zirconia oxide ion conductor based lambda (λ) probe relies on the electrostatic potential drop, which is built up at the junction between the metallic (Pt) electrode and the ceramic oxide. Only electrons may be exchanged across this interface, which alone build up the electrical field like in a Schottky barrier. In contrast to a common electronic device, the concentration of the electrons is not fixed by preparation but depends on the oxygen partial pressure. When the oxygen activity is lowered, the number of holes is decreasing and the number of electrons is increasing in the oxide, which is reflected by a change in the electrical field. The metallic electrode layer has to be porous in order to allow the gas to equilibrate with zirconia. The lambda probe may be therefore considered to be an electronic device rather than an ionic one [2].

We have shown earlier that the response time depends on the diffusion of oxide ions in the zirconia electrolyte underneath the platinum electrode layer [3]. The electrons are conducted at the interface along the platinum metal and are in this way not rate-determining. Any slow bulk diffusion of neutral oxygen into or out of the zirconia, caused by the equilibration of the electrolyte upon the change of the surrounding oxygen partial pressure, depends on the diffusion of the electronic minority charge carriers, but this has no influence on the formation of the electrostatic potential drop at the interface.

Also, ceramic type II [4, 5] electrochemical gas sensors are electronic devices. The difference is only due to the equilibration of the gas with the immobile component. According to Duhem-Margules' relationship, the variation in the activity of the immobile species causes a driving force and diffusion of the mobile ions inside the ceramic electrolyte. These are diffusing underneath the metallic electrode similarly to the case of the zirconia – platinum interface and cause in this way the electronic junction.

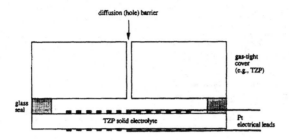

Fig. 1. Diffusion polarization controlled type I or type II amperometric chemical sensor with a barrier in the gas phase.

Only ceramic type III electrochemical gas sensors are based on the equilibration of the ions in addition to the electrons at the interface between the electrode and electrolyte. The electrode is a mixed conductor, which includes both the electro-active mobile component of the electrolyte and the gaseous species of interest [4 - 8]. The situation corresponds to that of batteries with the equilibration of the mobile species and electrons across the interface, but differs in such a way that the composition of the electrode depends on the gas partial pressure. In addition to the equilibration at the electrode-electrolyte interface, an equilibration at the surface of the electrode with a gas and within the electrode is required.

When a current is passed through the electrolyte, either by enforcing the current by the application of voltages from the outside or (partial or total) short-circuiting, the electro-active species have to be delivered to the ionic conductor at one side and taken off at the opposite electrode side. These processes require transport of the electro-active species toward and away from the electrolyte, electrochemical redox processes and transfer of the ions or atoms across the interfaces. Including the bulk resistance of the electrolyte, these processes are summarized as "polarizations". These kinetic phenomena depend on the kind of materials, i.e. chemical compositions and morphologies, and also on the kind of gaseous species. This offers an opportunity to take advantage of the kinetic interfacial processes for detecting and distinguishing the various components of gas atmospheres or fluids.

One of the various kinds of polarizations is or may be made rate-determining. This is applied in conventional amperometric sensors. In a very simple way, the diffusion of the interesting component towards the interface is controlled by a barrier built into the gas phase, as shown in Fig. 1. The same galvanic cells are being employed as in the case of potentiometric measurements. Type I and II amperometric sensors make use of ceramic ion conductors of the component which is being measured and immobile, but equilibrating with the mobile component, respectively. Type III amperometric sensors (Fig. 2) undergo a growth or decomposition of the auxiliary electrode layer by pumping the current and absorbing the species of interest [9, 10]. The variation in the thickness of the layer is commonly negligible at not too high concentrations of the detected species in the gas or fluid in view of the large amount of charges taken up per unit volume of solids. In addition, the layer may be rejuvenated by inversing the current. In all cases, the electrode reactions themselves are considered to be fast as compared to the transport of the atoms or molecules across the gas or fluid diffusion barrier.

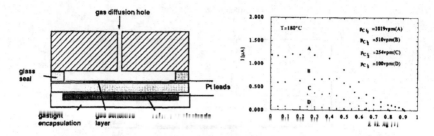

Fig. 2. Diffusion polarization controlled type III amperometric chemical sensor with a barrier in the gas phase. A reaction product (gas sensitive) layer is formed at the surface of the electrolyte between the electroactive component and the measured species. A current plateau is observed when the activity of the electroactive component is increased and accordingly the activity of the measured species is decreased.

Potentiometric gas sensors are considered to establish thermodynamic equilibria at the interfaces with the gas. For any voltage or EMF measurement it is required, however, to draw a (small) current. Even with the best available instruments of about 10^{15} Ω input resistance there are still approximately 10^4 species carrying charge passing the interface per second. If the concentration of species under detection becomes small in the gas or fluid phase, competing redox processes of other gaseous components may become predominant in spite of their less favorable chemical reactions. This is illustrated in Fig. 3 for the case of interaction of a type III gas sensor for chlorine gas which interferes with H₂S at approximately the same concentration of both gases in spite of the much lower stability of formation of Ag₂S [11, 12]. Another example is shown in Fig. 4, which indicates a lower chlorine activity in the presence of moisture by forming complex molecules with a water shell [13].

In the following, polarization processes immediately at the interfaces of the electrolyte and electrode will be discussed and considered for application in ceramic gas sensors.

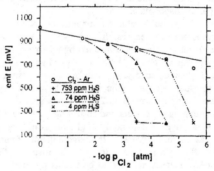

Fig. 3. Response of a potentiometric type III chlorine gas sensor in the presence of H₂S. In spite of the lower thermodynamic stability of Ag₂S compared to AgCl, the sulfide is being formed because of kinetic reasons when H₂S becomes predominantly present in the gas phase.

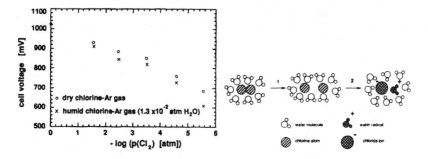

Fig. 4. Response of a potentiometric type III chlorine gas sensor in the presence of moisture. In this case, the chlorine activity is decreased compared to dry chlorine because of the formation of complex molecules with a water shell as illustrated at the right hand side.

3. PRINCIPLES OF THETA (Θ) SENSORS

Specific electrode reactions may be observed in many different ways for gases and fluids, e.g. by steady state Butler-Volmer relationships. However, it is more convenient for sensor applications to look for time dependent relationships, i.e. relaxation processes. One possibility is the application of periodical signals around the equilibrium voltage [14 – 17]. In this way, compositions of the various electrode phases are not being changed in the average over each cycle. This holds especially when the current is controlled, but is also approximately valid in the case of controlling the cell voltage.

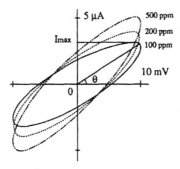

Fig. 5. Theta sensor arrangement and response to CO_2. Thin sodium carbonate layers have been employed at both sides of a ceramic Beta-Alumina solid sodium ion conducting electrolyte. The steady state relationship between the electrical current passed through the galvanic cell and the applied sinusoidal voltage is shown. The gas sensitive electrode may be in general formed also electrochemically in-situ. Furthermore, asymmetric electrodes may be employed.

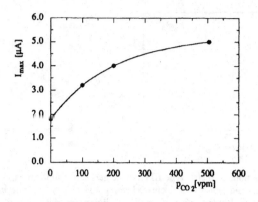

Fig. 6. Relationship between the maximum current of a θ sensor as derived from the curve shown in Fig. 6 and the CO_2 partial pressure. Similar monotonous dependencies exist for all frequencies and other characteristic quantities of the θ curves on the CO_2 partial pressure.

Conveniently, the current or voltage is varied sinusoidally and the dependant electrical quantity, i.e. voltage or current, respectively, is measured. The plot of the voltage versus current shows ellipsoidal figures with a hysteresis of the two electrical quantities. An example is shown in Fig. 5. A type III sensor with Beta-Alumina and a sodium carbonate layer on both sides of the ceramic electrolyte has been employed in this experiment. At one electrode side, the sodium diffuses into the electrode and interacts with the gas while the sodium is taken off from the sodium carbonate at the opposite side and CO_2 is liberated. It is clearly seen that the direction of the ellipsoidal figure changes typically with the CO_2 composition. In view of the shape of the curve, this type of sensor has been named theta (θ) sensor. The direction of the main axis, the maximum current and other characteristic features may be related to the gas partial pressure, as shown in Fig. 6 [18, 19].

The same principle may be employed in the case of the presence of several different gaseous species. Their interfacial reactions vary with regard to the rate determining steps and their concentrations. The different gases provide distinctive different fingerprints. A possibility to make these differences clearly visible is the Fourier analysis of the periodic response of the applied periodic electrical signal. The various "deformations" of the ellipsoidal figure by the interaction of the different gases with the interface become visible through the Fourier coefficients. Such an analysis has been made for the simultaneous detection of NO_2 and SO_2 [20]. The first order real part of the complex Fourier coefficient is plotted in Fig. 7 against the first order imaginary part in a so-called Argand diagram. Radial straight lines through the origin are observed for the case of the presence of only NO_2 in artificial air at concentrations from 150 – 3000 ppm (lower line) and various concentrations of SO_2 from 0.05 to 0.15 % together with 900 ppm NO_2. It is seen that the NO_2 and SO_2 gas concentrations determine the distance from the origin and the angle between the straight line through the experimental data point and the origin of the plot relative to the b_1 and α_1 axis. These two parameters indicate in this way in reverse the

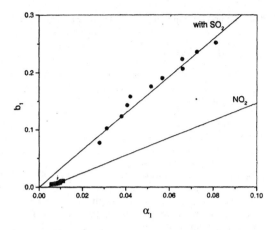

Fig. 7. Dependencies of the first order real and imaginary coefficient of the complex Fourier analysis on the partial pressures of NO_2 and SO_2. The lower straight line represents the presence of NO_2 alone while the upper straight line shows the results obtained with various SO_2 concentrations when simultaneously 900 ppm NO_2 was present. Both concentrations vary radially along straight lines through the origin. The influence of a second gas is indicated by a change in the angle of the straight line against either axis. Operation temperature: 550 °C; frequency: 50 mHz.

SO_2 and NO_2 concentrations separately. Both gas concentrations may be determined simultaneously within a period of time of the order of magnitude of the time required for one cycle of the applied signal. Suitable frequencies are in the range from 10 – 100 mHz, though this parameter has not been optimized yet. In case of more complex gas mixtures it may be necessary that higher Fourier coefficients have to be taken into consideration.

4. RATE-DETERMINING STEPS OF THE POLARIZATION

Besides the phenomenological description and verification of the practical applicability of the θ sensor concept, it is of interest to understand microscopically the electrode – gas or – fluid interactions. Therefore, the three polarization processes of diffusion of the interacting species toward and away from the interface, the adsorption process and the charge transfer are described in terms of their different time dependences, which will be compared with the experimental results. The three relationships are given by the following equations [20]:

i) Adsorption controlled current upon the application of a small sinusoidal voltage

$$I_{ads} = \frac{zF\omega V_0 Q}{2RT} \frac{\exp\left(-\frac{zFV_0}{2RT}\sin\omega t\right).\cos\omega t}{\left(1+\exp\left(-\frac{zFV_0}{2RT}\sin\omega t\right)\right)^2} \tag{1}$$

ii) Diffusion controlled current upon the application of a small sinusoidal voltage

$$I_{diff} = AzF\sqrt{\frac{D}{\pi}}c\left[\exp\left(\frac{zF}{RT}V_0\sin\omega t\right)-1\right]\frac{1}{\sqrt{t}} \qquad (2)$$

iii) Charge transfer controlled current (assuming a reversible transfer coefficient $\alpha = 0.5$) upon the application of a small sinusoidal voltage

$$I_{ch-tr} = 2AJ_0\left[\sinh\left(\frac{zF}{2RT}V_0\sin\omega t\right)\right] \qquad (3)$$

z, F, R, T, c, t, ω, V_0, J_0 and Q represent the charge number of mobile ions, Faraday's constant, general gas constant, absolute temperature, concentration, time, 2π times frequency of the applied periodic current or voltage signal, voltage amplitude. Exchange current density and amount of charge required to form a monolayer of adsorbed intermediates, respectively.

The quality of agreement with the experimental results is expressed by the mean error coefficient. The results are shown in Table 1 which indicate that the interaction of the sensor with SO_2 is controlled by diffusion toward and away from the interfaces. The polarization of NO_2 is determined by the charge-transfer reaction of the gas species at the ceramic electrolyte surface. This is probably a special situation to find different rate determining processes. Species with the same same rate determining process may become nevertheless detected and discriminated in the way described above since the electrically initiated driving forces provide different, specific responses as fingerprints.

Table I. Mean errors for three different potential rate determining kinetic steps for NO_2 and SO_2.

Step	$\sigma(NO_2)$	$\sigma(SO_2)$
Adsorption	0.111	0.433
Charge-transfer	0.030	0.339
Diffusion	0.241	0.084

Various types of electrodes may be employed. Since an electrical current will form the electro-active layer of type III sensors in-situ, there is in general no need to deposit such a layer separately beforehand. It may be also favorable to prepare the two electrodes differently and to have different rate determining processes for oxidation and reduction by different types of electrodes.

5. CONCLUSIONS

Conventionally, ceramic ionic conductors are being employed in chemical sensors in a passive way by observing the equilibrium cell voltage or by passing a current across the electrolyte. In both cases, steady state signals are generated which may depend on several gaseous species but may not become discriminated with regard to different types of chemical species. Additional information is needed. This may be provided by the described kinetic type of sensors, which are active devices by generating different modes of driving forces for the electrochemical reactions at the interfaces between the electrolyte, electrode and gas or fluid phase. There is a large choice of different signals, especially varying the amplitude of the periodic electrical quantity, frequency and shape of the time dependence in addition to different electrode chemistry and morphology, to achieve the best performance and highest sensitivity and selectivity of the sensors. Since we do not need to establish thermodynamic equilibria, it is also possible to apply the principle of θ sensors to conventional sensors, which do not establish equilibria because the temperatures are too low. Compared to "chess board" type multiple sensor arrays, the advantage of the θ principle is obviously the application of only one single galvanic cell element for the simultaneous detection of a multiple number of components. As long as there is no net transport of electrical charge across the ionic conductor, the device does not change chemically with time and long life times are expected.

6. ACKNOWLEDGMENT

The presented work has been the result of the contributions of several coworkers, especially S. Bredikhin, W.F. Chu, P. Jasinski, J. Liu, A. Menne, F. Salam, E. Steudel, E.D. Tsagarakis. Their support is gratefully appreciated.

7. REFERENCES

[1] Applications of Solid Electrolytes (T. Takahashi and A. Kozawa, Eds.), JEC Press, Cleveland, Ohio, 1980
[2] W. Weppner, Ionics 7, 404 - 424 (2001)
[3] E.D. Tsagarakis and W. Weppner, in: Solid-State Ionic Devices: Ceramic Sensors (E.D. Wachsman, W. Weppner, E. Traversa, M. Liu, P. Vanysek and N. Yamazoe, Eds.), Proc. Vol. 2000-32, Electrochem. Soc., Pennington, NJ, 2001, pp. 285 – 297
[4] W. Weppner, in: Proc. 2nd Int. Meet. Chem. Sensors (J.-L. Aucouturier, J.-S. Cauhapé, M. Destrian, P. Hagenmuller, C. Lucat, F. Ménil, J. Portier and J. Salardienne, Eds.), Bordeaux, France, 1986, pp. 59 - 68; Sensors and Actuators 12, 107 - 119 (1987)
[5] W. Weppner, in: Electrochemical Detection Techniques in the Applied Biosciences. Vol. 2 - Fermentation and Bioprocess Control, Hygiene and Environmental Sciences (G.A. Junter, Ed.), Ellis Horwood, Chichester, GB, 1988, pp. 142 - 160
[6] W.F. Chu, E.D. Tsagarakis, T. Metzing and W. Weppner, Ionics 9, 321 – 328 (2003)
[7] F. Salam and W. Weppner, Ionics 4, 355 – 359 (1998)
[8] W. Weppner, Materials Science and Engineering B 15, 48 - 55 (1992)
[9] J. Liu and W. Weppner, Appl. Phys. A 52, 94 - 99 (1991)
[10] A. Menne and W. Weppner, in: Proceedings Third International Meeting on Chemical Sensors, Cleveland, Ohio, USA, Sept. 24-26, 1990, pp. 225 – 228
[11] J. Liu and W. Weppner, Sensors and Actuators B 6, 270 - 273 (1992)
[12] A. Menne and W. Weppner, Electrochim. Acta 36, 1823 - 1827 (1991)
[13] A. Menne and W. Weppner, Sensors and Actuators B 9, 79 - 82 (1992)
[14] J. Liu and W. Weppner, Applied Physics A 55, 250 - 257 (1992)
[15] E. Steudel and W. Weppner, in: Solid-State Ionic Devices (E.D. Wachsman, J.R. Akridge, M. Liu and N. Yamazoe, Eds.), Proc. Vol. 99-13, The Electrochemical Society, Inc., Pennington, NJ, 1999, pp. 310 – 323

[16] P. Jasinski, A. Nowakowski and W. Weppner, Materials and Sensors 12, 89 – 97 (2000)
[17] E.D. Tsagarakis, W.F. Chu, T. Metzing and W. Weppner, in: Solid-State Ionic Devices: Ceramic Sensors (E.D. Wachsman, W. Weppner, E. Traversa, M. Liu, P. Vanysek and N. Yamazoe, Eds.), Proc. Vol. 2000-32, Electrochem. Soc., Pennington, NJ, 2001, pp. 270 – 284
[18] B.Y. Liaw, J. Liu, A. Menne and W. Weppner, Solid State Ionics 53 - 56, 18 - 23 (1992)
[19] S. Bredikhin, J. Liu and W. Weppner, Appl. Phys. A 57, 37 - 43 (1993)
[20] E.D. Tsagarakis, Investigation of Kinetic Processes of Gas-Solid-Conductor-Interfaces with respect to Potential Application in Chemical Sensors, Ph.D. Thesis, Faculty of Engineering, Christian-Albrechts University, Kiel, Germany, 2001

METAL-OXIDE BASED TOXIC GAS SENSORS

Duk Dong Lee
School of Electronic Engineering & Computer Science, Kyungpook National University
1370, Sankyuk Dong, Pukgu,
Daegu, 702-701

Nak Jin Choi
Future Technology Research Division, Electronics and Telecommunications Research Institute
161 Gajeong Dong, Yusenggu,
Daejoenog, 305-350

ABSTRACT
Recently, there has been increasing demand for monitoring the highly toxic gases known as chemical warfare agents such as nerve, vesicant, blood, choking agents. For this purpose, metal oxide semiconductor gas sensors have been investigated and shown very promising results.

In this study, thick film SnO_2-based gas sensors added with Al_2O_3, In_2O_3, ZnO, and ZrO_2 using alumina substrate were fabricated, and their material properties and sensing properties for the four different simulant chemical warfare agents of dimethyl methyl phosphonate, dipropylene glycol methyl ether, acetonitrile, dichloromethane have been examined. A micro sensor array with 6 devices using silicon substrate was designed and fabricated using MEMS technology to realize high sensitive and high selective sensing system with low power consumption and provide the portable chemical agents detecting system.

The thick film sensing devices exhibited quite high sensitivity to the toxic gases at ppm level. And the micro sensor using MEMS technology showed high sensitivity at the device temperature of 300 °C which could be obtained by only 90 mW heater power consumption.

Key Words: Tin oxide, Thick film, MEMS, Chemical warfare agents, Si process

I. INTRODUCTION

Major progresses have been made over the last few years as regards understanding the mode of action and treatment of classical, biological, and chemical warfare agents. CWA is divided into four types of blood, nerve, vesicant, and choking agent [1,2]. CWA is very dangerous for health because of its colorlessness and toxicity. Therefore, the fast and accurate detection of CWAs is essential to protect human beings. Many researchers have already attempted to detect the gas by utilizing various kinds of sensors including metal oxide semiconductor sensors (MOSs), quartz-crystal microbalance sensors (QCMs), electrochemical sensors, and surface acoustic wave sensors (SAWs), etc [3-6]. Among the above-mentioned sensors, SnO_2-based semiconductor sensors have many advantages, such as high sensitivity, fast response, good reproducibility and good long-term stability, furthermore they are hardly affected by ambient disturbances, at least compared with other kinds of sensor [7]. The structures of semiconductor gas sensors which are in common use can be divided into two categories; thick and thin films [8-10]. In general, thick film sensor exhibits a high sensitivity at low concentrations, due to the large specific surface available for chemisorption reactions. However, lack of stability is one of its drawbacks. On the other hand, the thin film sensor has a high stability and repeatability, but lower sensitivity [11,12].

In this study, tin oxide based thick films were formed on both alumina and silicon substrate using screen printing method. To reduce the power consumption, the thick films were formed on the insulation layer membrane obtained by backside etching of insulation and silicon layer [13]. The sensing materials prepared are characterized by using thermogravimetric analysis (TGA), X-ray diffraction (XRD), scanning electron microscopy (SEM), and specific surface area analyzer (BET: Brunaure, Emmett, Teller). The response properties of the sensors are examined for test gases. Several less poison gases are used as simulating CWA: methyl phosphonate (DMMP) as a stimulant of Tabun (nerve agent gas), dipropylene glycol methyl ether (DPGME) as a stimulant of Nitrogen mustard (vesicant agent gas), acetonitrile (CH_3CN) as a simulant of HCN (blood agent gas), and dichloromethane (CH_2Cl_2) as a simulant of Phosgene (choking agent gas) [1].

II. EXPERIMENTS
2.1. Sensor Preparation
2.1.1 Fabrication of single sensor on alumina substrate

Fig.1 shows the schematic view of single sensor on alumina substrate. Single sensor was used to analyze sensing materials and to examine their responses for four stimulant gases. Semiconductor sensors need a heater due to the requirement of high thermal energy in the gas response. The sensor consists of one heater and a pair of sensing electrodes. Pt sensing electrode is deposited with 1000 Å thickness using DC sputter at the front side of alumina substrate and the heater was screen printed using the Pt paste at the back side. Then device is thermally annealed at 850 °C for 10 min in the electric furnace. The heater resistance is 10 Ω and sensing electrode is designed as interdigitated (IDT) structure. Overall dimension of the device was $7 \times 10 \times 0.6$ mm^3 and size of sensing film is 6×4 mm^2.

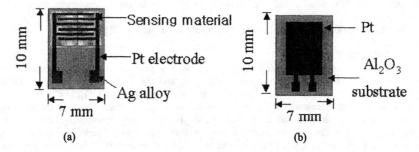

(a)   (b)

Fig. 1. Schematic view of single sensor using alumina substrate. (a) Front view (b) back view.

2.1.2 Fabrication of micro sensor array using Si substrate

Fig. 2 shows the schematic view of sensor array on Si substrate. A micro sensor array was designed and fabricated to reduce the power consumption and to provide high selectivity to the gas.

A fabrication flowchart for the micro gas sensor is shown in Fig. 3. A SiNx thin film with thickness of 2 μm was fabricated using the low pressure chemical vapor deposition (LPCVD) process as a membrane on a 4-inch 100 p-type wafer with a 500 μm thickness. A Ta thin film

26

with a 300 Å thickness and Pt thin film with a 2000 Å thickness were then deposited by a sputtering process, followed by the spin coating of a photoresist film of AZ1512 with a 1.2 µm thickness that was then patterned using a mask and ultra violet lithography. Thereafter, reactive ion etching (RIE) process was carried out to etch the Pt film to form a Pt heater and temperature sensors, except for the photoresist-coated region, then passivation layers to protect against conduction between the heater and the sensing electrode were formed based on SiO_2 films using the plasma enhanced chemical vapor deposition (PECVD) process. The deposition pressure and temperature were 3.8 mTorr and 400 °C, respectively, and the RF power and deposition rate were 200 W and 125 Å/sec, respectively. A photoresist film with a 1.2 µm thickness was spin coated for the opening pad area and patterned using a mask and ultra violet lithography. The SiO_2 film was then selectively etched using RIE, except for the photoresist-coated region. Also, the SiNx thin film on the back-side was selectively etched using the same process mentioned above for the membrane formation after silicon bulk micromachining. The heater resistance is 50 Ω and the sensing electrode is designed as interdigitated (IDT) structure. Overall dimension of micro array sensor was $8 \times 12 \times 0.5$ mm^3 and the size of sensing area fabricated was 1×1 mm^2.

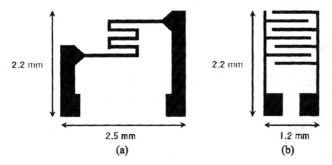

Fig. 2. Schematic view of micro sensor array using Si substrate. (a) Heater electrode (b) sensing electrode.

2.2. Preparation of Sensing Materials

Table 1 lists the gas sensing materials used in this study. Base material for the of all samples is SnO_2. The sensing materials are prepared by following two different processes. 1) Al_2O_3 (0, 4, 12, 20 wt.%) and In_2O_3 (1, 2, 3 wt.%) were added to SnO_2 by physical ball milling process [12]. 2) ZnO (1, 2, 3, 4, 5 wt.%) and ZrO_2 (1, 3, 5 wt.%) were added to SnO_2 by co-precipitation method. All samples are dried at 110 °C for 1 h, followed by grinding and calcination at 600 °C for 1 h in the furnace. Ethylcellulose and organic binder (α-terpineol) were added into the powder to make slurry, powder was deposited on SiNx surface using screen-printing technique. After dring, all sensors were further sintered at 700 °C for 1 hr. The fabricated sensors were then tested after allowing them to stabilize at 400 °C for 3 days.

2.3. Apparatus and Measurement Method

Table 2 lists the measurement gases. Typical chemical agents are shown in Table 2 [1]. To use the real chemical gases is very dangerous and difficult, so simulant gases with similar chemical characteristics are used as test gases for safety reason. Test gases are DMMP, DPGME, CH_3CN, and CH_2Cl_2.

The four simulant agents are liquid phase, therefore the vapor partial pressures are calculated to control the injected gas concentration. Acetonitrile and dichloromethane adapt Antoine equation as shown in Equation (1) and DMMP uses Equation (2). In the case of DPGME, the vapor pressures were 0.3 and 0.05 mmHg at 20 and 25 °C, respectively [14].

$$\log_{10}P = A - B/(T+C) \quad \text{-------------------------(1)}$$

$$P = 2.844 \times 10^8 \times \exp(-11500/RT) \quad \text{------------------(2)}$$

Where, P (mmHg) means vapor pressure, the constants A, B, and C are specific values of each chemical. T(K) is gas temperature. As shown in Equation (1) and (2), gas concentration can be controlled by changing the circulator (device to sustain the temperature) temperature.

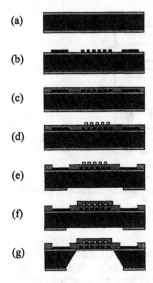

Fig. 3. Fabrication flowchart for micro gas sensor. (a) Depositing of Si_3N_4 with 2 μm thickness layer by LPCVD process (b) Depositing of Pt 2000 Å heater, photoresist film of AZ1512 and outside etching of Pt heater by RIE (mask #1) (c) Depositing of SiO_2 insulation layer by PECVD (d) Depositing of Pt 2000 Å electrode, photoresist film of AZ1512 and outside etching of Pt electrode by RIE (mask #2) (e) Opening of electrode using RIE (mask #3) and patterning of backside etching area (mask #4) (f) Depositing of sensing materials (SnO_2 with Al_2O_3), and (g) Etching of backside of Si using KOH

Table 1. Amounts of added materials and material synthesis method.

Added material	Added amounts (wt.%)	Synthesis method
Al_2O_3	0, 4, 12, 20	Ball-milling
In_2O_3	1, 2, 3	Ball-milling
ZnO	1, 2, 3, 4, 5	Co-precipitation
ZrO_2	1, 3, 5	Co-precipitation

Table 2. Measurement simulant gases [1,12].

Agent	Chemical gas	Simulant agent
Blood	HCN	Acetonitrile
Nerve	Tabun	DMMP (dimethyl methyl phosphonate)
Choking	CE	Dichloromethane
Vesicant	HN	DPGME (dipropylene glycolmethyl ether)

* CE and HN mean Phosgene and Nitrogen mustard respectively [1].

Table 3 shows specific values in the Antoine equation. T range means that this equation can be used in the range of the temperature. Fig. 4 shows measurement apparatus. Sample gas with various concentrations can be prepared by using mass flow controllers. The total mass flow fixed at 1000 ml/min. Circulator fixes the atmosphere temperature of reactant. The sensing signal, the sensing film resistance, was acquired using a data acquisition board (E6024 NI Co., USA.) [15], which was simultaneously able to acquire 16 channels of analog input and scan speed in this experiment was 10 samples per second.

Table 3. Values of used parameters at equation 1.

	Acetonitrile	Dichloromethane
T range(℃)	7-23	-40-40
A	5.93296	4.53691
B	2345.829	1327.016
C	43.815	-20.474

III. RESULTS

3.1. Characteristics of Sensing Materials

The fabricated sensing materials were characterized by TG-DTA (TA instrument, U.S.A.) for the selection of calcination temperature, SEM (Hitachi Co., Japan) for their surface morphology and thickness, XRD (Rigaku Co., Japan) for their particle size, and BET for their specific surface area. The calcination temperature was decided as 600 °C from the results of TG-DTA analysis. The surface morphology of the materials was found to be comparatively uniform [16], and the thickness of the sensing layer was observed approximately 20 μm by SEM analysis. In the result of XRD analysis, large peaks of all fabricated materials coincide with SnO_2 peaks of JCPDS card. Peaks of additive materials were not found in the XRD analysis because added amounts were very small [16]. Table 4 shows the summary of powder characteristics for the four sensing materials by various analyses. The crystallite sizes were calculated from Scherrer's equation [16]. The crystallite sizes of the materials fabricated by ball-milling method (SnO_2-Al_2O_3, SnO_2-In_2O_3) were between 25 to 46 nm and their specific surface areas were between 6 to 9 m^2/g. But the crystallite sizes and the specific areas of materials prepared by co-precipitation method (SnO_2-ZnO, SnO_2-ZrO_2) were between 4 to 17 nm and between 20 to 70 m^2/g, respectively.

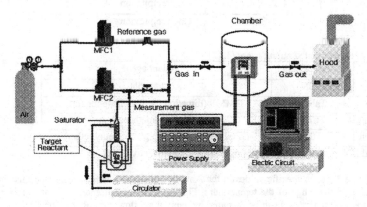

Fig. 4. Measurement apparatus.

Table 4. The characterization of the four sensing materials.

Materials	Crystallite size(nm)	Specific surface area (m^2/g)	Pore size (nm)	Optimal operating temperature (°C)
SnO_2-Al_2O_3	35	7.5	32	250-350
SnO_2-In_2O_3	45	7.5	35	300-400
SnO_2-ZnO	12	45	40	200-400
SnO_2-ZrO_2	4	23	35	200-400

3.2. Single Sensor and Micro Sensor Array

Fig. 5 shows the photographs of the fabricated single sensor and the micro sensor array. Fig. 5(a), (b), and (c) are the front view, backside view, and complete sensor which is printed by a screen-printing method, in turns. Fig. 5(d), (e), and (f) are the front view, backside view after backside etching, and complete sensor which is printed by a screen-printing method, in turns. A sensor array was consisted of 6 heaters and 6 sensing electrode pairs.

Fig. 6 shows temperature variation vs. power consumption for the single sensor and a micro sensor, one of the 6 sensors. Fig. 6(a) is power consumption graph of single sensor using alumina substrate and Fig. 6(b) is that of micro sensor using Si substrate. This graph shows real heater temperature as applied power changes, by infrared thermometer (IR308, Minolta Co. Japan). This instrument can measure from 250 to 800 °C, and target spot size is 1.3 mm at 100 mm distance. The temperature of the heater linearly increased in proportion to applied power. At the operating temperature of 300 °C, the respective power consumption was 3000 mW for alumina substrate sensor and 90 mW for Si substrate based sensor.

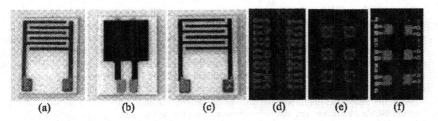

Fig. 5. Photographs of single sensor by alumina substrate and micro sensor array by Si substrate. (a) Sensing electrode(front side) (b) heater(backside) (c) sensor screened (d) front side (e) backside view after backside etching (f) Sensor screened.

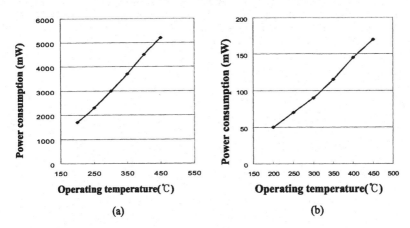

Fig. 6. Power consumption vs. operating temperature. (a) Alumina substrate (b) Si substrate

3.3. Responses of Single Sensor for Four Simulants

If the gas being detected is adsorbed in the form of positive charges, electrons become trapped on the semiconductor surface, thereby moving the conduction band, so the resistance of the semiconductor is decreased. Conversely, the adsorption of a negatively-charged gas/odor, like NO_2, means the resistance of the semiconductor is increased. The sensitivity is defined as follows.

$$S (\%) = (Rg-Ra)/Ra \times 100 \quad \text{------------------}(3)$$

where Ra and Rg represent the resistance in air atmosphere and that in gas ambient, respectively.

In the present study, the sensor with different added amounts of additive materials were characterized at different operating temperatures to select the appropriate amounts of additive materials and operating temperature. Fig. 7 shows graphs of the additive material amounts vs. the sensitivity at different operating temperatures, where the test gases were DMMP for Al_2O_3 and

31

In$_2$O$_3$, and DPGME for ZnO and ZrO$_2$, and the gas concentration was fixed at 0.5 ppm. As shown in the figures, SnO$_2$ with the addition of 4 wt.% Al$_2$O$_3$ exhibited the highest sensitivity, and then decreased above 4 wt.%. Also, the highest sensitivity was exhibited at 300 °C. In general, when the operating temperature is increased, the gas desorption on the surface becomes faster. Semiconductor gas sensor has specific temperature with the higher sensitivity. As such, the sensitivity was proportional to the temperature increase until 300 °C. However, the sensitivity became inversely proportional to temperature increases above 300 °C. From the figures, added amounts of four materials were selected to deposit on sensing electrode of Si substrate. The recommendable additives and the ratio of mixing with base material were 4wt.%-Al$_2$O$_3$, 2wt.%-In$_2$O$_3$, 2wt.%-ZnO and 1wt.%-ZrO$_2$.

Fig. 8 shows a real-time graph of the PC output. The sensing film SnO$_2$(98 wt.%)-ZnO(2 wt.%) and operating temperature 300 °C were fixed. After stabilizing for 30 min., a gas flow meter was used to inject the test gas for 10 min, followed by an injection of fresh air for 15 min at 10 times. The injected concentration was fixed DPGME 0.5 ppm. As shown in Fig. 8, the repetition test was excellent, below 5 % on a full scale.

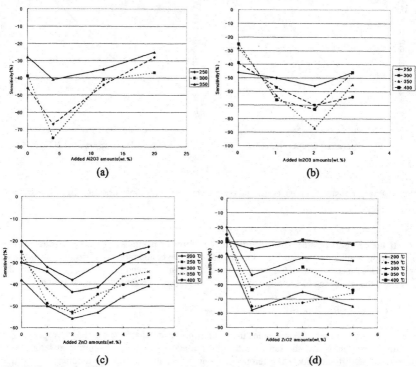

(a) (b)

(c) (d)

Fig. 7. Graphs of additive amount vs. sensitivity for four sensing materials at different operating temperatures. (a) SnO$_2$-Al$_2$O$_3$ (b) SnO$_2$-In$_2$O$_3$ (c) SnO$_2$-ZnO (d) SnO$_2$-ZrO$_2$

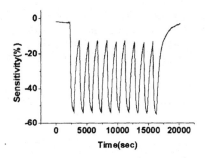

Fig. 8. Repetition test for 0.5 ppm DPGME at 300 ℃.

3.4. Responses of Micro Sensor Array for Four Simulants

Fig. 9 shows the sensitivity vs. gas concentrations of four materials to four stimulant agents. In Fig. 9(a), up to 15 ppm of four agents, the sensitivity steeply increased, but over 15 ppm, the change moderately increased. All four materials show like that.

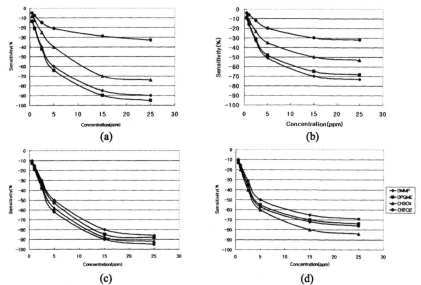

Fig. 9. Sensitivity vs. gas concentrations of four materials. (a) Al_2O_3 (b) In_2O_3 (c) ZnO (d) ZrO_2

Sensitivity graphs at different operating temperatures were shown in Fig. 10. Operating temperature was changed from 250 to 350 °C. Injection concentration was fixed at 2.5 ppm of four stimulant agents. Responses were changed with operating temperature. For example, the shapes of CH3CN and CH2Cl2 changed as temperature increase. At 250 °C, Al2O3 and In2O3 dopped materials show low sensitivity for CH3CN and CH2Cl2. But, at 350 °C, they show comparatively high sensitivity for CH3CN and CH2Cl2. By using these results, simulant agents can be discriminated by using sensor array.

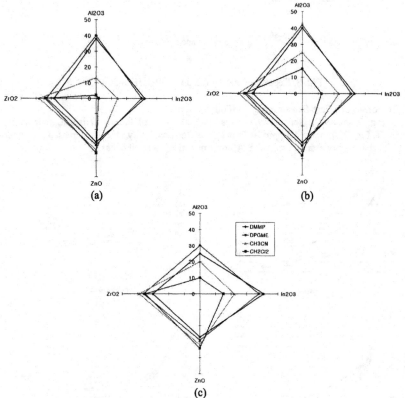

Fig. 10. Response to test gases with different operating temperatures (2.5 ppm).
(a) 250 °C   (b) 300 °C   (c) 350 °C

PCA results are shown in Fig. 11. PCA is known as one of the statistical distribution methods. For example, if we assume four parameters, the parameters are shown four dimensions in the graph. That is difficult to understand at a glance. But PCA presents two or three dimensions by extracting significant data from various parameters [10]. Therefore the graph by PCA method was shown very simple. PCA is adapted to classify the chemical agent. Fig. 11 shows the classification among the various gases. Test gases are four simulant agent and CH4,

34

C_4H_{10}, H_2, and CO that are toxic gases which can co-exist with chemical agents. In Fig. 11, numbers such as 0.5, 3 mean gas concentrations injected. And the principal component (PC) value, 98.9, indicates importance of specific axis out of whole axis. In other words, assume if sum of whole axis importance was 100, 98.9 number means that X-axis occupies about 99. Although two points in the X-axis exist very near, two points can be classified as other class. As shown in figure, test gases can be easily classified by using a sensor array.

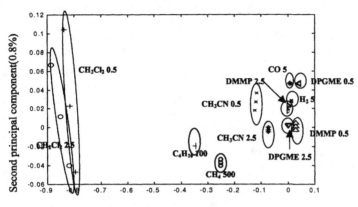

First principal component(98.9%)

Fig. 11. Classification among other gases.

IV. CONCLUSIONS

Micro sensor array with thick film based dopped SnO_2 on tin oxide was fabricated on a Si substrate and examined its characteristics were examined for the chemical agents. The sensor dimensions were $8 \times 12 \times 0.5$ mm^3, and Al_2O_3, In_2O_3, ZnO, and ZrO_2 were incorporated into SnO_2 to enhance the sensitivity and selectivity to the specific chemical agent. The responses to four simulant gases were examined for different gas concentrations and different amounts of additives. The repetition measurements were performed to see the signal variation was within in the test, it is said to be 5 %. A sensor array with micro devices which were added with different additives was fabricated and PCA was adapted to classify target gases.

Consequently, a sensor array fabricated by using thick film and MEMS technology showed high sensitivity to simulant gases, operated with low power consumption and could discriminate between gases by using the PCA method.

ACKNOWLEDGEMENTS

The authors acknowledge the financial support from the Chemical and Biological Terror Confrontation Program and National Research Laboratory Program of the Korean Ministry of Science and Technology.

REFERENCES
[1]T. C. Marrs, R. L. Maynard and F. R. Sidell, *Chemical warfare agents : Toxicology and treatment*, John Wiley & Sons, New York, 1-5, (1996).
[2]J. C. Lee, "Chemical weapons destruction technology (I)," *J. Korea Solid Wastes Engineering Society*, 16, 207, (1999).

[3]J. W. Gardner, "Detection of vapours and odours from a multisensor array using pattern recognition Part 1. Principal component and cluster analysis," *Sens. Actuators B*, 4, 109 -115, (1991).
[4]T. Nakamoto, A. Fukuda, "Perfume and flavour identification by odour-sensing system using quartz-resonator sensor array and neural-network pattern recognition," *Sens. Actuators B*, 10, 85-90, (1993).
[5]H. Nanto, T. Kawai, H. Sokooshi, T. Usuda, "Aroma identification using a quartz-resonator sensor in conjunction with pattern recognition," *Sens. Actuators B*, 14, 718-720, (1993).
[6]T. J. Lee, H. Y. Song and D. J. Chung, "ICP etching of Pt thin films for fabrication of SAW devices," *J. Korean Phys. Soc.*, 42, 814-818, (2003).
[7]Toru Maekawa, Kengo Suzuki, Tadashi Takada, Tetsuhiko Kobayashi, Makoto Egashira, "Odor identification using a SnO_2-based sensor array," *Sens. Actuators B*, 80, 51-58, (2001).
[8]D. S. Lee, D. D. Lee, "NO_2 sensing properties of WO_3-based thin film gas sensors," *J. Korean Phys. Soc.*, 35, 1092-96, (1995).
[9]Y. S. Lee, B. S. Joo, N. J. Choi, B. H. Kang and D. D. Lee, "Pattern recognition of a gas sensor array using impedance," *J. Korean Phys. Soc.*, 37, 862-865, (2000).
[10]N. J. Choi, C. H. Shim, K. D. Song, B. S. Joo, J. K. Jung, O. S. Kwon, Y. S. Kim and D. D. Lee, "Recognition of indoor environmental gases using two sensing films on a substrate," *J. Korean Phys. Soc.*, 41, 1058-62, (2002).
[11]W. Y. Chung, "Tungsten oxide thin films prepared for NO_2 sensors by using the hydrothermal method and dip coating," *J. Korean Phys. Soc.*, 41, 181-183, (2002).
[12]N. J. Choi, B. T. Ban, J. H. Kwak, W. W. Baek, J. C. Kim, J. S. Huh and D. D. Lee, "Fabrication of SnO_2 gas sensor added by metal oxide for DMMP," *J. Korea Institute of Military Science and Technology*, 6, 54-61, (2003).
[13]J. W. Lim, D. W. Kang, D. S. Lee, J. S. Huh, and D. D. Lee, "Heating power-controlled micro-gas sensor array," *Sens. Actuators B*, 77, 139-144, (2001).
[14]NIST chemistry WebBook, (2003).
[15]D. Y. Kwak, *LabVIEW^{TM} Control of computer based and measurement solution*, Ohm, Seoul, 229-250, (2002).
[16]N. J. Choi, T. H. Ban, J. H. Kwak, W. W. Baek, J. C. Kim, J. S. Huh and D. D. Lee, "Fabrication of DMMP thick film gas sensor based on SnO_2," *J. Korean Institute of Electrical and Electronic Material Engineers*, 16, 1217-23 (2003).

THERMALLY STABLE MESOPOROUS SnO$_2$ AND TiO$_2$ POWDERS FOR SEMI-CONDUCTOR GAS SENSOR APPLICATION

Yasuhiro SHIMIZU and MAKOTO EGASHIRA*
Graduate School of Science Technology, *Faculty of Engineering
Nagasaki University, 1-14 Bunkyo-machi, Nagasaki 852-8521, JAPAN

ABSTRACT
        Mesoporous and large mesoporous SnO$_2$ and mesoprous TiO$_2$ powders were prepared and then their potentials as semiconductor gas sensor materials have been examined. Phosphoric acid treatment of as-prepared powders was found to be effective for suppressing the crystallite growth and then maintaining their ordered mesoporous and large mesoporous structure as well as large specific surface area up to elevated temperatures. Owing to their thermally stable ordered porous structure, mesoporous and large mesoporous sensors showed larger gas response than the sensors fabricated with conventionally prepared and commercially available powders. However, the improvement in response was relatively smaller than expected from the enlargement in specific surface area. Formation of large secondary particles during the aging of the synthesized products and poisoning of active sites for gas detection, especially for mesoprous TiO$_2$, are considered to be possible reasons for the unexpectedly small response enhancement. Simultaneous surface modification of conventional SnO$_2$ powder, i.e. loading of Ru and subsequent coating with a mesoporous SnO$_2$, was proved to be an effective approach in improving gas sensing properties. The improved response was considered to arise from a synergistic effect of the diffusion control by the mesoporous layer and the chemical sensitization by the Ru loaded.

INTRODUCTION
        A particular attention is currently being given to nanosized or nanostructured semiconductor metal oxides as sensor materials[1,2], since grain-size reduction[3] and gas-diffusion control[4,5] have been proved to be useful for improving the sensing properties, especially for the magnitude of gas response, of semiconductor gas sensors. After the successful preparation of mesoporous silica[6], which is characterized with controlled mesoporous structure, nanosized crystallite and large specific surface area, numerous efforts have been directed to preparing mesoporous semiconductor metal oxides by employing a similar preparation procedure, i.e. by employing the self-assembly of supramolecules. Besides the application to catalyst supports, these mesoporous semiconductor metal oxides are of technological interest as gas sensor materials from the viewpoints of gas-diffusion control as well as the increased active sites for gas detection, in comparison with those prepared by conventional methods. However, poor thermal stability of mesoporous structure of SnO$_2$, which is a typical gas sensor material, synthesized so far[7,8] limits its application to gas sensors, which are usually operated in the temperature range of 200-500°C.
        Thus, our recent studies have been directed to developing thermally stable mesoporous and large mesoporous SnO$_2$ and mesoporous TiO$_2$ (denoted as m-SnO$_2$, lm-SnO$_2$ and m-TiO$_2$, respectively) powders and then to evaluating their potentials as semiconductor gas sensor materials. Although TiO$_2$ is not a common semiconductor gas sensor material due to its intrinsically poor sensing properties, i.e. low reactivity of oxygen adsorbates on the TiO$_2$ surface with inflammable gases, enlargement of specific surface area was considered to make this

material more attractive as a semiconductor gas sensor. Further attempts were made to evaluate the effects of simultaneous surface modification of the SnO_2 powder, which was prepared by a conventional method, with an m-SnO_2 layer and noble metal loading on the gas sensing properties.

EXPERIMENTAL

Preparation of Thermally Stable m-SnO_2 and lm-SnO_2

Mesoporous SnO_2 powder was prepared by employing $Na_2SnO_3 \cdot 3H_2O$ as an Sn-source and n-cetylpyridinium chloride ($C_{16}PyCl$) as a surfactant under the following conditions[9]: $C_{16}PyCl$ concentration in deionized water = 2 wt%, $[C_{16}PyCl]/[Na_2SnO_3 \cdot 3H_2O]$ = 2 (in molar ratio), pH = 10, aging at 20°C for 3 days. Mesitylene (trimethylbenzene, MES) was added to the precursor solution at a molar ratio $[MES]/[Na_2SnO_3 \cdot 3H_2O]$ = 2.5 in some cases. The resultant solid product was allowed to stand for 3 days in the solution at 20°C. Filtered products were additionally treated in a phosphoric acid (PA, 0.1 M) solution for about 2 h to enhance their thermal stabilities. The products were then calcined at 600°C for 5 h in air. Mesoporous SnO_2 powder prepared without the PA treatment is simply expressed as m-SnO_2, while PA treated m-SnO_2 powders prepared with and without the MES addition are referred to as PA/m-SnO_2 and PA/m-SnO_2(MES), respectively. Hereafter, the powder calcined at 600°C will be expressed such as PA/m6-SnO_2(MES), where the numeral 6 represents the calcination temperature of 600°C. For reference, another kind of SnO_2 powder (c6-SnO_2) was prepared by a conventional method, that is, calcination of tin oxalate at 600°C for 5 h in air.

Large mesoporous SnO_2 powder was prepared in a manner similar to that for m-SnO_2, but by employing $SnCl_2 \cdot H_2O$ as an Sn-source, a triblock poly(ethylene oxide)-b-poly(propylene oxide)-b-poly(ethylene oxide) copolymer (P123, $EO_{20}PO_{70}EO_{20}$, mean molecular weight: 5800) as a surfactant and 2 M urea as a pH adjuster.[10] Powders were obtained after the aging at 80°C for 24 h of the precursor solutions under the following conditions: concentration of P123 was varied in the range of 0.67–2.1 wt%, but the molar ratio of $[P123]/[SnCl_2 \cdot H_2O]$ was fixed at 0.034, unless otherwise noted. Powders prepared with and without PA treatment (0.1 N, at RT for 2 h) were subjected to heat treatment at 600°C for 5 h in air. As-prepared large mesoporous powders thus prepared with and without the PA treatment are referred to as lm-SnO_2 and PA/lm-SnO_2, in a manner similar to the above, while calcined powders are expressed as lm6-SnO_2 and PA/lm6-SnO_2, respectively.

Preparation of Thermally Stable m-TiO_2

Mesoporous TiO_2 powder was prepared by employing a modified sol-gel method with $Ti(NO_3)_4$ and polyethylene glycol having an average molecular weight of 6000 (PEG6000) in a manner similar to that reported by Liu et al.[11] During the preparation, precipitated gel was subjected to PA treatment (0.1 M, at RT for 2 h) in some cases, prior to heat treatment at 100°C for 3 h, wherein the hydrolysis and condensation took place within the PEG6000 matrix. Mesoporous TiO_2 with and without the PA treatment are also referred to as m-TiO_2 and PA/m-TiO_2. Then the resulting powder was calcined at 500°C for 1 h in air.[12] The calcined powders are expressed as m5-TiO_2 and PA/m5-TiO_2, respectively.

Simultaneous Surface Modification of c-SnO_2

Conventionally prepared c6-SnO_2 powder was used as a base sensor material in investigating the effect of simultaneous surface modification with coating of a m-SnO_2(MES) layer and noble metal loading.[13]   Loading of 0.5 wt% Ru or Pd on the c-SnO$_2$ powder was conducted by a conventional method: suspending c6-SnO$_2$ powder in an aqueous solution dissolving RuCl$_3$ or PdCl$_2$, subsequent evaporation to dryness, followed by treatment under flowing H$_2$ at 350°C for 5 h.   The surface of unloaded and metal-loaded c-SnO$_2$ powders was modified with a m-SnO$_2$(MES) layer in a manner similar to that reported previously.[14]   The powders subjected to the surface modification for two times were also prepared.   The resultant products were subjected to PA treatment (0.1 N, at RT for 2 h) and then were calcined at 600°C for 5 h in air.   The powders thus prepared are denoted as {PA/m6-SnO$_2$(MES)(n)}/c6-SnO$_2$ or {PA/m6-SnO$_2$(MES)(n)}/{0.5Ru (or 0.5Pd)/c6-SnO$_2$}, where n represents the number of times of the surface modification with the PA/m6-SnO$_2$(MES) layer and 0.5 stands for the loading amount of the metal in weight percent, respectively.   Loading of 0.5 wt% Ru or Pd on {PA/m6 -SnO$_2$(MES)(2)}/c6-SnO$_2$ powder was further conducted in the same manner, and the resultant powder was denoted as 0.5Ru (or 0.5Pd)/{PA/m6-SnO$_2$(MES)(2)}/c6-SnO$_2$.

Characterization and Measurement of Sensing Properties
        Crystal phase and crystallite size (CS) of the powders were characterized with X-ray diffraction analysis (XRD; Rigaku, RINT2200).   The CS value was calculated by Scherrer's equation.   The specific surface area (SSA) and pore size distribution were measured by the BET method using a N$_2$ sorption isotherm (Micromeritics, TriStar3000).   Morphology of the powders and resulting thick films was observed by a scanning electron microscope (SEM; Hitachi, S-2250N) and a transmission electron microscope (TEM; JEOL, JEM2010-HT).
        Thick film sensors were fabricated by applying the paste of the SnO$_2$-based powders on alumina substrates having a pair of interdigitated Pt electrodes (the gap between electrodes: 200 μm), followed by calcination at 500°C (for c6-SnO$_2$ and m-SnO$_2$ series powders) or 600°C (for lm-SnO$_2$ and simultaneously surface modified c6-SnO$_2$ series powders) for 5 h in air.   In the case of the TiO$_2$-based powders, however, disk-type sensors were fabricated aiming at reducing the sensor resistance in air to a practical level.   The TiO$_2$ powders were pressed into disks (5.0 mmφ, 0.5 mm thick) and were calcined at 550°C for 3 h in air.   Palladium paste was applied (in ca. 3.0 mm in diameter) on both surfaces of the disk and then fired at 550°C for 1 h in air to serve as electrodes.   Gas response of the sensors was measured to a sample gas balanced with air in a flow apparatus in a given temperature range.   The magnitude of gas response (k) was defined as the ratio (R_a/R_g) of sensor resistance in air (R_a) to that in the sample gas balanced with air (R_g).

RESULTS AND DISCUSSION
Characterization of Thermally Stable m-SnO$_2$ and Its Sensing Properties
        The usefulness of the PA treatment and the MES addition in enhancing the thermal stability can be confirmed by the XRD patterns of the resultant m-SnO$_2$ powders before and after calcination, as shown in Fig. 1.   The appearance of a peak around $2\theta = 2°$, assigned to a 100 diffraction peak, reveals that all the as-prepared powders are in ordered mesoporous structure with a d_{100} value, which corresponds to the distance between ordered SnO$_2$ layers, of about 4 nm. However, it is obvious that the ordered mesoporous structure of m-SnO$_2$ is fractured completely by calcination at 600°C for 5 h in air, as shown in Fig. 1(b), indicating poor thermal stability of m-SnO$_2$.   In the case of m-SnO$_2$ powder, the CS of SnO$_2$ increased markedly from 1.9 to 9.4 nm by the calcination.   Thus, the growth of SnO$_2$ crystallites is considered to be responsible for the fracture of the ordered mesoporous structure.   In contrast, the PA treatment is confirmed to be

effective for improving the thermal stability of the ordered mesoporous structure, since the 100 diffraction peak appears even after calcination at 600°C, as shown in Fig. 1(d). But, the diffraction peak shifts to a larger angle by the calcination, corresponding to a decrease in d_{100}-value from 4.1 to 3.2 nm. Furthermore, the addition of MES in the precursor solution is found to highlight the effect of the PA treatment in improving the thermal stability, since a higher 100 peak appears after the calcination as shown in Fig. 1(f). As for the PA/m6-SnO$_2$ and PA/m6-SnO$_2$(MES) powders, no change and only a slight change from 1.7 to 2.0 nm in CS, respectively, was induced by the calcination. Thus, the PA treatment is confirmed to be useful for reducing the growth of SnO$_2$ crystallites and then improving the thermal stability of the ordered mesoprous structure.

Irrespective of the decrease in the 100 peak intensity, its shift to a higher diffraction angle and the disappearance of the 110 diffraction peak, ordered mesoprous structure with ca. 1.6 nm spacing is maintained in almost all regions of PA/m6-SnO$_2$(MES) grains even after calcination at 600°C for 5 h in air, as shown in Fig. 2. In addition, thickness of each SnO$_2$ layer is also found to be about 1.6 nm. Therefore, these results are in good agreement with the value of d_{100}, i.e. 3.1 nm. Furthermore, the 1.6 nm spacing is almost the same as the pore diameter at the maximum pore volume measured for PA/m6-SnO$_2$(MES), as shown in Fig. 3(b). Owing to such ordered mesoporous structure, PA/m6-SnO$_2$(MES) has a large SSA of 305 m^2 g^{-1} even after the calcination. This value was much larger than those of calcined m6-SnO$_2$ (40.5 m^2 g^{-1}) and c6-SnO$_2$ (8.4 m^2 g^{-1}), and then PA/m6-SnO$_2$(MES) showed larger pore volume than these two powders, as shown in Fig. 3. From these results, PA/m-SnO$_2$(MES) was confirmed to have sufficient thermal stability for gas sensor application and certain improvement in gas sensing properties was expected from its unique ordered mesoporous structure.

Figure 4 shows operating temperature dependence of thick film PA/m6-SnO$_2$(MES) and c6-SnO$_2$ sensors to 500 ppm H$_2$ balanced with air. The sensor thickness was ca. 85 μm for PA/m6-SnO$_2$(MES) and 100 μm for c6-SnO$_2$. It is obvious that the PA/m6-SnO$_2$(MES) sensor exhibits larger H$_2$ response than the c6-SnO$_2$ sensor over the whole temperature range studied, e.g. sixfold response at 350°C. Thus, the results shown in Fig. 4 demonstrated clearly the potential of the thermally stable ordered mesoporous SnO$_2$ powder as a semiconductor gas sensor material. But, the H$_2$ response enhancement induced by the introduction of the ordered mesoporous structure was not so remarkable in comparison with that expected from the increase in SSA of the sensor material: only sixfold response in spite of 36-fold SSA. Possible reasons for such unexpectedly small response of the PA/m6-SnO$_2$(MES) sensor will be discussed below.

Characterization of Thermally Stable lm-SnO$_2$ and Its Sensing Properties

Also in the case of lm-SnO$_2$ powders, the usefulness of the PA treatment in improving their thermal stabilities has been confirmed by comparing the XRD patterns of the powders calcined at elevated temperatures, as shown in Fig. 5. Here, lm-SnO$_2$ series powders were prepared under a P123 concentration of 2.1 wt% and a [P123]/[SnCl$_2$·H$_2$O] molar ratio of 0.036, while 2.0 wt% and 0.034, respectively, for PA/lm-SnO$_2$ series powders. As-prepared lm-SnO$_2$ powder was amorphous, but the degree of crystallinity increased obviously with a rise in calcination temperature, and then diffraction peaks of a tetragonal SnO$_2$ phase appeared clearly after calcination at 600°C for 5 h. The CS value increased accordingly from 7.6 nm for lm4-SnO$_2$ to 12 nm for lm6-SnO$_2$. In contrast, diffraction peaks of PA/lm6-SnO$_2$ remained broad even after calciantion at 600°C for 5 h, as shown in Fig. 5(e). In addition, the change in CS upon calcination at 600°C was very small, i.e. from 3.4 nm to 4.0 nm. Thus, the PA

40

treatment was confirmed again to be effective for suppressing the growth of crystallites and then maintaining ordered large mesoporous structure up to elevated temperatures. The formation of ordered and large mesoporous structure was confirmed by the pore size distribution data shown in Fig. 6. Namely, the PA/lm6-SnO$_2$ powder showed a sharp pore size distribution with a peak maximum around 4 nm and a SSA of 142 m^2 g^{-1}, as shown in Fig. 6(d). The larger pore size observed for PA/lm6-SnO$_2$ than PA/m6-SnO$_2$(MES) is undoubtedly ascribed to a larger molecular size of P123 used as a surfactant. Pore volume and then SSA of PA/lm6-SnO$_2$ varied with the P123 concentrations in the precursor solutions. Figure 7 shows pore size distributions of a series of PA/lm6-SnO$_2$ powders prepared under different P123 concentrations, but at the fixed [P123]/[SnCl$_2$·H$_2$O] molar ratio of 0.034. The data shown in Fig. 6(d) are cited again in Fig. 7(a) for easy comparison. The pore diameter at the maximum pore volume remained unchanged, irrespective of the change in the P123 concentration. But, the largest pore volume and in turn the largest SSA value of 253 m^2 g^{-1} were achieved at a P123 concentration of 1.3 wt%. Hereafter, PA/lm6-SnO$_2$ series powders are distinguished from each other with a labeling of P123 concentration, e.g., PA/lm6-SnO$_2$(2.0-P123) for the P123 concentration of 2.0 wt%.

Figure 8 shows operating temperature dependence of response of several thick film lm-SnO$_2$ sensors to 1000 ppm H$_2$ balanced with air. Thickness of these sensors was controlled to be around 30 μm. Among the sensors tested, PA/lm6-SnO$_2$(2.0-P123) showed the largest H$_2$ response at 400°C. However, the difference in H$_2$ response among the sensors was very small, irrespective of a large change in SSA of sensor materials. Relationship between the logarithmic H$_2$ response and the SSA values of the sensors is depicted in Fig. 9. For comparative purpose, Fig. 9 also shows the data obtained with several thick film PA/m6-SnO$_2$(MES) sensors prepared with different C$_{16}$PyCl concentrations as well as neutralization conditions and with another thick film SnO$_2$ sensor fabricated with the powder prepared by calcination of tin oxalate at 900°C for 2 h (denoted as c9-SnO$_2$) in our previous studies.[15] It is obvious that all PA/lm6-SnO$_2$ sensors showed relatively large H$_2$ response, despite the small SSA values, in comparison with those for the PA/m6-SnO$_2$(MES) sensors. However, no clear correlation existed between the SSA values and the H$_2$ response. This result may imply that the inner SnO$_2$ surface of mesoporous and large mesoporous structure is not so active for H$_2$ detection, due to the limitation of gas diffusion into the mesoporous structure, even for the PA/lm6-SnO$_2$ series sensors having a pore diameter of ca. 4 nm.

Another notable feature of m-SnO$_2$ and lm-SnO$_2$ powders is formation of large secondary particles in the range of 2-5 μm in diameter just after the precipitation, although the size of secondary particles remained almost unchanged upon calcination up to 600°C. This presents a striking contrast to the case of the c-SnO$_2$ powder consisting of fine particles less than submicron in size even after calcination at elevated temperatures. As a typical example, an SEM photograph of the PA/m6-SnO$_2$(MES) sensor is compared with that of the c6-SnO$_2$ sensor in Fig. 10. The appearance of the PA/lm6-SnO$_2$ sensor resembled to that of the PA/m6-SnO$_2$(MES) sensor, and the c9-SnO$_2$ sensor also consisted of very fine particles as the c6-SnO$_2$ sensor, though photographs were not shown here. In the cases of the m-SnO$_2$ and lm-SnO$_2$ series sensors, therefore, it is considered that the H$_2$ response is determined mainly by the surface region of large secondary particles, especially at grain-boundaries among the secondary particles. Small numbers of grain-boundaries inside the m-SnO$_2$ and lm-SnO$_2$ series sensors is, therefore, considered to be responsible partly for the smaller response. Other possibilities are poisoning of the active SnO$_2$ surface by the PA treatment and existence of more important factors determining sensor response other than the morphological characteristics, such as SSA, grain size, secondary grain size and sensor thickness, though the details are not clear at present.

Characterization of Thermally Stable m-TiO₂ and Its Sensing Properties

Formation of ordered mesoporous structure in the as-prepared powders (m-TiO$_2$ and PA/m-TiO$_2$) was confirmed by the appearance of a small diffraction peak around 2.1°, as shown in Figs. 11(a) and 11(c). Then, these powders showed a large SSA of more than 200 m^2 g^{-1} and consisted of small crystallites of ca. 5 nm. Without the PA treatment, the calcination at 500°C resulted in disappearance of the diffraction peak around 2.1°, a decrease in SSA value from 214 to 25.5 m^2 g^{-1} and an increase in CS from 4.4 to ca. 15 nm. Such microstructural changes imply fracture of the ordered mesoporous structure by the calcination. But, a large and clear diffraction peak around 2.1° remained after the calcination by the PA treatment, as shown in Fig. 11(d). In addition, the calcination caused little changes in both SSA and CS and then no phase transition. Thus, the suppression of both the crystallite growth and the partial phase transformation from anatase to rutile induced by the PA treatment is responsible for the improvement of the thermal stability of the ordered mesoporous structure.

Figure 12 shows operating temperature dependence of response of m5-TiO$_2$ and PA/m5-TiO$_2$ sensors to 500 ppm H$_2$ and 500 ppm CO balanced with air. For comparative purpose, the results obtained with another disk-type sensor which was fabricated from commercially available TiO$_2$ powder (SSA: 9.7 m^2 g^{-1}, CS: 200 nm) are also depicted in Fig. 12. Since the commercially available powder was used as received, i.e. without any calcination treatments prior to the sensor fabrication, the sensor is expressed simply as c-TiO$_2$. The PA/m5-TiO$_2$ sensor showed larger H$_2$ and CO response than the c-TiO$_2$ sensor, indicating usefulness of the introduction of the mesoporous structure in TiO$_2$ powder in improving its gas sensing properties. But, again the response was not improved significantly as was expected from the enlargement of SSA. In addition, SEM observation confirmed the formation of large secondary particles also in the cases of ordered mesoporous TiO$_2$ powders, while c-TiO$_2$ powder consisted of fine particles less than submicron in diameter.[16] Thus, it is revealed that both the ordered mesoporous SnO$_2$ and TiO$_2$ powders tend to form large secondary particles under the present preparation conditions even before being subjected to subsequent calcination. More excellent sensing properties may be achieved by modifying the preparation conditions so as to get smaller particles in future. Another important feature to be noticed is again the influence of the PA treatment on the sensing properties. The PA treatment was confirmed to be effective for improving the thermal stability of the ordered mesoporous structure also in the case of TiO$_2$ powder. However, the PA/m5-TiO$_2$ sensor showed smaller response to both H$_2$ and CO than the m5-TiO$_2$ sensor, irrespective of its ordered mesoporous structure along with large SSA, while selectivity to H$_2$ of the PA/m5-TiO$_2$ sensor was superior to the m-TiO$_2$ sensor. At least, therefore, the PA treatment obviously deteriorated the sensing ability of m-TiO$_2$ powder. Thus, development of an alternative method for improving the thermal stability is of primary importance to realize further enhancement of gas sensing properties of ordered mesoporous semiconductor metal oxides.

Characterization of Simultaneously Surface Modified c-SnO₂ and Its Sensing Properties

Pore size distribution of c6-SnO$_2$ and {PA/m6-SnO$_2$(MES)(n)}/c6-SnO$_2$ are shown in Fig. 13. SSA of the c6-SnO$_2$ powder used in this subject was 9.54 m^2 g^{-1} and was slightly larger than that used in the above subject, probable due to different preparation batches. SSA as well

42

as pore volume at a pore diameter of ca. 3 nm increased with increasing the number of surface modification cycles with a PA/m6-SnO$_2$(MES) layer. Although the pore diameter at the maximum pore volume for {PA/m6-SnO$_2$(MES)(2)}/c6-SnO$_2$ became larger than that for the PA/m6-SnO$_2$(MES) powder (see the data of Fig. 3(b)), the results shown in Fig. 13 clearly demonstrated successful deposition or formation of a PA/m6-SnO$_2$(MES) layer on the surface of c6-SnO$_2$ particles by the modification. In addition, thickness or area of the PA/m6-SnO$_2$(MES) layer deposited was confirmed to increase with increasing the number of surface modification cycles by TEM observation, as shown in Fig. 14.

Figure 15 shows operating temperature dependence of H$_2$ response of sensors fabricated with the c6-SnO$_2$-based powders either modified with PA/m6-SnO$_2$(MES) or loaded with a noble metal. Thickness of these sensors was controlled to be in the range of 30-40 μm. It is obvious that the H$_2$ response improved gradually with increasing the number of the surface modification cycles over the whole temperature range studied. Thus, the PA/m6-SnO$_2$(MES) layer deposited is considered to act as a diffusion control layer[4,5], especially for gaseous oxygen, due to the difference in molecular size between O$_2$ and H$_2$. This induces relatively higher H$_2$ concentration in a very small space between the PA/m6-SnO$_2$(MES) layer and the surface of c6-SnO$_2$ particles or grain-boundaries than in the surrounding atmosphere, and then to larger response. This scenario is, of course, based on the assumption that the surface of c6-SnO$_2$ is more active for H$_2$ detection than PA/m6-SnO$_2$(MES) surface, at least from the viewpoint of extracting electrical signals, i.e. resistance changes, from the sensor. The validity of this assumption will be discussed by referring to additional data in the below. Loading of 0.5 wt% Ru or Pd on the c6-SnO$_2$ powder was also found to be effective for the improvement of the response, especially at lower temperatures less than 400°C. The improvement induced by these metals undoubtedly arises from the chemical sensitization effect[17]: acceleration of chemical reactions between H$_2$ and chemisorbed oxygen on the c6-SnO$_2$ powder surface and then achievement of a new but low surface coverage of the chemisorbed oxygen at the steady-state in the H$_2$ atmosphere, in comparison with that in the case of no noble metal loading.

More significant improvement in H$_2$ response could be achieved with the twice surface modification of 0.5Ru/SnO$_2$ with the PA/m6-SnO$_2$(MES) layer, i.e. {PA/m6-SnO$_2$(MES)(2)} /{0.5Ru/c6-SnO$_2$}, as shown in Fig. 16. In contrast, no significant change in response was induced by the loading of 0.5 wt% Ru on the surface of {PA/m6-SnO$_2$(MES)(2)}/c6-SnO$_2$, i.e. 0.5Ru/{PA/m6-SnO$_2$(MES)(2)}/c6-SnO$_2$, with an exception of a slight improvement at an operating temperature of 350°C. This result confirms that the PA/m6-SnO$_2$(MES) layer is no longer active for H$_2$ detection, in comparison with the c6-SnO$_2$ powder surface, in the case of the {PA/m6-SnO$_2$(MES)(2)}/c6-SnO$_2$ sensor. Thus, the improvement in H$_2$ response observed with {PA/m6-SnO$_2$(MES)(2)}/c6-SnO$_2$ can be concluded to arise from the diffusion control of gaseous oxygen by the PA/m6-SnO$_2$(MES) layer. In addition, the markedly improved H$_2$ response of the {PA/m6-SnO$_2$(MES)(2)}/{0.5Ru/c6-SnO$_2$} sensor is considered to arise from a synergistic effect of the diffusion control by the PA/m6-SnO$_2$(MES) layer and the chemical sensitization by the loaded Ru on the c6-SnO$_2$ powder surface. In the case of 0.5Ru/{PA/m6-SnO$_2$(MES)(2)}/ c6-SnO$_2$, the loaded Ru is considered to limit the permeation amount of H$_2$ into the SnO$_2$ powder surface by the accelerated reaction between H$_2$ and chemisorbed oxygen on the PA/m6-SnO$_2$(MES) layer surface. This weakens the sensitization effect by the PA/m6-SnO$_2$(MES) layer, and then leads to a H$_2$ response property similar to that for {PA/m6-SnO$_2$(MES)(2)} /c6-SnO$_2$.

In the case of the Pd loading before and after the surface modification with the

PA/m6-SnO$_2$(MES) layer, on the other hand, the H$_2$ response decreased slightly on the whole, in comparison with {PA/m6-SnO$_2$(MES)(2)}/c6-SnO$_2$, as shown in Fig. 17, although the H$_2$ response of {PA/m6-SnO$_2$(MES)(2)}/{0.5Pd/c6-SnO$_2$} at temperatures higher than 450°C was slightly larger than {PA/m6-SnO$_2$(MES)(2)}/c6-SnO$_2$. Reduction of the permeation amount of H$_2$ into the SnO$_2$ powder surface is considered to become significant when Pd is loaded after the surface modification with the PA/m6-SnO$_2$(MES) layer, due to higher catalytic activity of Pd than Ru. Therefore, the decrease in H$_2$ response in the case of 0.5Pd/{PA/m6-SnO$_2$(MES)(2)} /c6-SnO$_2$, in comparison with {PA/m6-SnO$_2$(MES)(2)}/c6-SnO$_2$, can be explained by this phenomenon. However, the fact that {PA/m6-SnO$_2$(MES)(2)}/{0.5Pd/c6-SnO$_2$} showed H$_2$ response comparable to {PA/m6-SnO$_2$(MES)(2)}/c6-SnO$_2$ raises another important factor in determining the response properties. The possible factor is diffusion behavior of a sample gas throughout the thick film sensor, i.e. its permeation behavior into the innermost region of the sensor atop the interdigitated electrodes. In the case of thick film sensors, it has been proved that inflammable gases are likely consumed at the surface region of the thick films when the catalytic activity of the sensor materials is high enough and then this leads to smaller response in the interior region of the thick films than the surface region[4,5], which is expressed as a configurational effect[18], even if the thick film acts as a gas diffusion control layer for gaseous oxygen. The sensor structure employed in the present study is a thick film type having electrodes in its innermost region. Thus, such a phenomenon likely occurs more or less for all the sensors tested. However, H$_2$ response of the Pd loaded sensors, i.e. 0.5Pd/{PA /m6-SnO$_2$(MES)(2)}/c6-SnO$_2$ and {PA/m6-SnO$_2$(MES)(2)}/{0.5Pd/c6-SnO$_2$}, is considered to be affected more significantly by this effect, due to higher catalytic activity of Pd than Ru. Furthermore, the reason for the comparable response of {PA/m6-SnO$_2$(MES)(2)} /{0.5Pd/c6-SnO$_2$} to {PA/m6-SnO$_2$(MES)(2)}/c6-SnO$_2$ may be ascribed to the configurational effect. Anyway, the results shown here demonstrate the potential of the simultaneous modification, i.e. the coating with a mesoporous layer and the loading of a noble metal having an appropriate catalytic activity, in improving the gas sensing properties of conventional semiconductor gas sensor materials.

CONCLUSION

Introduction of thermally stable ordered mesoporous and large mesoporous structure into semiconductor metal oxides has been proved to be effective for improving their gas sensing ability to some extend. More excellent sensing properties may be achieved by modifying preparation conditions so as to get smaller particles of mesoprous and large mesoporous powders as sensor materials. Besides the phosphoric acid treatment, development of an alternative method for improving the thermal stability is of primary importance to realize further excellent sensing properties. Furthermore, simultaneous surface modification with loading of a noble metal having an appropriate catalytic activity and subsequent coating of a mesoporous layer was proved to be an effective approach in improving gas sensing properties of conventionally prepared SnO$_2$ power. This method can be applied to other conventional semiconductor gas sensor materials and is considered to be also useful for the improvement of their sensing abilities.

REFERENCES
[1]A.M. Ruiz, A. Cornet and J. R. Morante, "Study of La and Cu Influence on the Growth Inhibition and Phase Transformation of Nano-TiO$_2$ Used for Gas Sensors," *Sens. Actuators B*, **100**, 256-60 (2004).

[2]C. Baratto, G. Sberveglieri, A. Onischuk, B. Caruso and S. di Stasio, "Low Temperature Selective NO_2 Sensors by Nanostructures Fibers of ZnO," *Sens. Actuators B*, **100**, 261-65 (2004).

[3]C. Xu, J. Takaki, N. Miura, and N. Yamazoe, "Relationship Between Gas Sensitivity and Microstructure of Porous SnO_2," *Denki Kagaku (Presently, Electrochemistry)*, **58**, 1143-48 (1990).

[4]Y. Shimizu, Y. Nakamura, and M. Egashira, "Effects of Diffusivity of Hydrogen and Oxygen Through Pores of Thick Film SnO_2-based Sensors on Their Sensing Properties," *Sens. Actuators B*, **13/14**, 128-31 (1993).

[5]Y. Shimizu, T. Maekawa, Y. Nakamura, and M. Egashira, Effects of Gas Diffusivity and Reactivity on Sensing Properties of Thick Film SnO_2-based Sensors, *Sens. Actuators B*, **46**, 163-68 (1998).

[6]C.T. Kresge, M.E. Leonomicz, W.J. Roth, J.C. Vartuli and J.S. Beck, "Ordered Mesoporous Molecular Sieves Synthesized by a Liquid-crystal Template Mechanism," *Nature*, **359**, 710-12 (1992).

[7]N. Ulagappan and C.N.R. Rao, "Mesoporous Phases Based on SnO_2 and TiO_2," *Chem. Commun.*, **14**, 1685-86 (1996).

[8]K.G. Severin, T.M. Abdel-Fattah and T.J. Pinnavaia, "Supramolucular Assembly of Mesostructured Tin Oxide," *Chem. Commun.*, **14**, 1471-72 (1998).

[9]T. Hyodo, N. Nishida, Y. Shimizu and M. Egashira, "Preparation and Gas-Sensing Properties of Thermally Stable Mesoporous SnO_2," *Sens. Actuators B*, **83**, 209-15 (2002).

[10]Y. Shimizu, A. Jono, T. Hyodo and M. Egashira, "Preparation of Large Mesoporous SnO_2 Powder for Gas Sensor Application," *Sens. Actuators B*, in press.

[11]X. Liu, J. Yang, L. Wang, X. Yang, L. Lu and X. Wang, "An Improvement on Sol-gel Method for Preparing Ultrafine and Crystallized Titania Powder," *Mater. Sci. Eng.*, **A289**, 241-45 (2000).

[12]C. Yu, T. Hyodo, Y. Shimizu and M. Egashira, "Preparation of Thermally Stable Mesoporous TiO_2 Powder and Its Gas Sensor Application," *Electrochemistry*, **71**, 475-80 (2003).

[13]Y. Shimizu, K. Tsumura, T. Hyodo and M. Egashira, "Effect of Simultaneous Modification with Metal Loading and Mesoporous Layer on H_2 Sensing Properties of SnO_2 Thick Film Sensors", *Trans. Inst. Elect. Eng. Jpn.*, in press.

[14]T. Hyodo, S. Abe, Y. Shimizu and M. Egashira, "Gas Sensing Properties of Ordered Mesoporous SnO_2 and Effects of Coatings Thereof," *Sens. Actuators B*, **93**, 590-600 (2003).

[15]T. Hyodo, Y. Shimizu and M. Egashira, "Design of Mesoporous Oxides as Semiconductor Gas Sensor Materials", *Electrochemistry*, **71**, 387-393 (2003).

[16]G. S. Devi, T. Hyodo, Y. Shimizu, M. Egashira, "Synthesis of Mesoporous TiO_2-based Powders and Their Gas-sensing Properties," *Sens. Actuators B*, **87**, 122-129 (2002).

[17]N. Yamazoe and N. Miura, "Some Basic Aspect of Semiconductor Gas Sensors," *in Chemical Sensor Technology*, S. Yamauchi, Editor, Vol.4, 19-41, Kodansha-Elsevier, New York (1992).

[18]Y. Shimizu and M. Egashira, "Basic Aspects and Challenges of Semiconductor Gas Sensors," *MRS Bulletin*, **24**, 18-24 (1999).

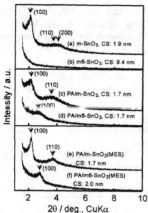

Fig. 1 XRD patterns of several m-SnO$_2$ series powders.

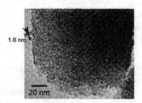

Fig. 2 TEM photograph of PA/m6-SnO$_2$(MES).

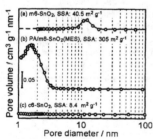

Fig. 3 Pore size distributions of several SnO$_2$-based powders.

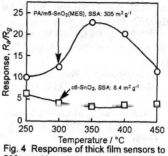

Fig. 4 Response of thick film sensors to 500 ppm H$_2$.

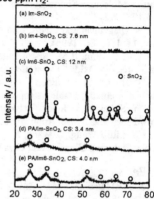

Fig. 5 XRD patterns of several Im-SnO$_2$ series powders.

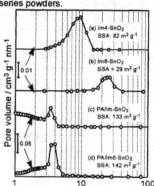

Fig. 6 Pore size distributions of several Im-SnO$_2$ series powders.

46

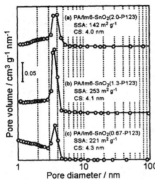

Fig. 7 Pore size distributions of several PA/Im6-SnO₂ powders prepared under different P123 concentrations.

Fig. 10 SEM photographs of thick film PA/m6-SnO₂(MES) and c6-SnO₂ sensors.

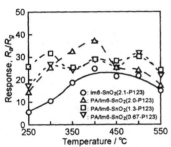

Fig. 8 Response of thick film sensors to 1000 ppm H₂.

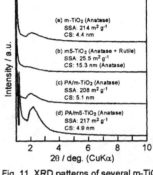

Fig. 11 XRD patterns of several m-TiO₂ series powders.

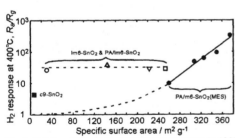

Fig. 9 Relationship between logarithmic response to 1000 ppm H₂ at 400°C and specific surface area of thick film SnO₂-based sensors.

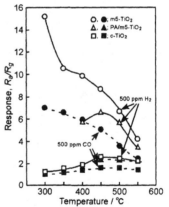

Fig. 12 Response of disk-type sensors to 500 ppm H₂ and 500 ppm CO

47

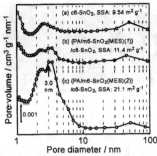

Fig. 13 Pore size distributions of c6-SnO₂ powders with and without surface modification by PA/m6-SnO₂ (MES).

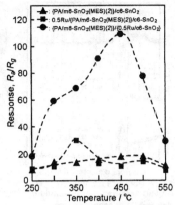

Fig. 16 Response of thick film sensors to 1000 ppm H₂.

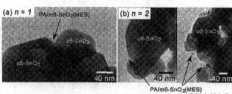

Fig. 14 TEM photographs of {PA/m6-SnO₂(MES)(*n*)}/c6-SnO₂ powders.

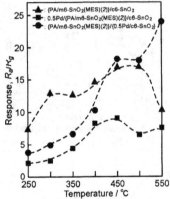

Fig. 17 Response of thick film sensors to 1000 ppm H₂.

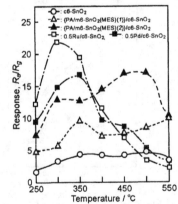

Fig. 15 Response of thick film sensors to 1000 ppm H₂.

DC ELECTRICAL-BIASED, ALL-OXIDE NO$_x$ SENSING ELEMENTS FOR USE AT 873 K

David West, Frederick Montgomery, and Timothy Armstrong
Oak Ridge National Laboratory
Oak Ridge, TN 37831-6083

ABSTRACT

All-oxide NO$_x$ sensing elements based on a solid electrolyte are reported. During operation, a DC current or voltage is applied to the element, and the resultant voltage or current is used as a sensing signal. The elements are nearly "total NO$_x$.", with responses to NO and NO$_2$ of the same algebraic sign and similar magnitude. For example, 77 ppm$_V$ NO produces a voltage change of -37% and 77 ppm$_V$ NO$_2$ a change of -40% for a current-biased element at 873 K and 7 vol% O$_2$. The elements display an [O$_2$] sensitivity that is a decreasing function of [NO$_x$] and their DC electrical resistance varies exponentially with temperature. These sensing elements could find application in sensors for measuring [NO$_x$] at temperatures near 900 K.

INTRODUCTION

Environmental stewardship is placing increasing demands on emission control of vehicle exhausts. For spark ignited, direct injection (SIDI) engine exhausts, these demands are largely met with the use of a "three-way catalyst" (TWC), capable of simultaneous remediation of CO, NO$_x$, and hydrocarbons. However, current generation TWC's lose their effectiveness for NO$_x$ reduction in the presence of excess O$_2$[1] and, therefore, cannot be applied to NO$_x$ remediation of the O$_2$-rich exhausts from diesel and lean-burn gasoline engines. Two technologies proposed for NO$_x$ remediation of these exhausts are the lean-NO$_x$ trap (LNT) and selective catalytic reduction (SCR). Both of these approaches will require on-board NO$_x$ sensors, to control trap regeneration with the use of LNT's or reagent injection in the case of SCR.

The "NO$_x$" in combustion exhausts is typically a mixture of NO and NO$_2$. The monoxide is the dominant equilibrium form at T > 500 °C (Fig. 1), but equilibrium conditions cannot be assumed. If [NO]/[NO$_2$] is unknown, it may be useful to have a "total NO$_x$" sensor with equal response characteristics to either NO or NO$_2$. Such a sensor would measure [NO] + [NO$_2$].

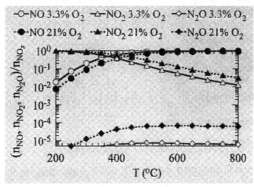

Figure 1: Equilibrium amounts of NO$_x$ species in mixtures of N$_2$ and O$_2$.[2]

Kato et al.[3] achieved near total NO_x sensing by lowering $[O_2]$ to ~1000 ppm and operating at 700 °C, thus driving the reaction $NO_2 = NO + \frac{1}{2}O_2$ far to the right. Szabo and Dutta[4] described the use of an upstream catalyst to fix the ratio $[NO]/[NO_2]$ prior to impingement on a mixed-potential NO_x sensing element. Nair et al.[5] employed a similar strategy to yield total NO_x sensing. Finally, Kunimoto et al.[6] employed "electrochemical NO_x conversion" (to NO_2) in a total NO_x sensor.

These prior efforts all incorporated some form of NO_x conversion or equilibration in the design of a total NO_x sensor. Here we describe our efforts to produce total NO_x *sensing elements*, sensing elements that respond equally to NO and NO_2 without the need for NO_x conversion or equilibration. High-T (700 °C) AC-biased sensing elements with this behavior have been reported by Miura et al.[7] but, in contrast to that work, the sensing elements reported here function with a DC bias signal and have all-oxide electrodes.

EXPERIMENTAL

As shown in Fig. 2a, the sensing elements consist of interdigitated oxide electrodes on a yttria-stabilized zirconia (YSZ) substrate (diameter ~1.6 cm, thickness ~0.1 cm). The YSZ substrates are fabricated in-house from Tosoh TZ8YS powder by tape casting, lamination, and sintering at 1400 °C for 2 hr. The electrodes are patterned onto the YSZ substrate by screen-printing dispersions at a wet print thickness of ~75 μm, drying at 130 °C, then firing at 1200 °C for 0.5 hr. Non-contact profilometry, as well as cross-sectioning and examination with the optical microscope, indicate that typical dried and fired electrode thicknesses are ~25 μm.

To evaluate sensing performance, pressure contacts (Pt wire) are made to the electrodes and gas mixtures of N_2, O_2, and either NO or NO_2 are presented to the electroded side of the sensing elements as shown schematically in Fig. 2b. An Environics 4000 mixing unit is used to mix the gases and the flow rate typically employed is 750 sccm. The element operating temperature is monitored by a type K thermocouple (not shown in Fig. 2b) placed about 1 cm from the element surface.

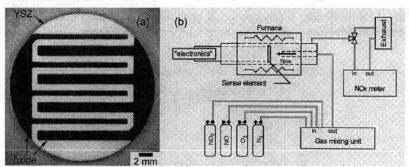

Figure 2: Sensing element (a) and schematic of test setup (b).

The "electronics" indicated in Fig. 2b differ depending on the goal of any particular investigation. If the voltage (with no external bias) between the electrodes is of interest, a Keithley 617 electrometer is used to measure the voltage. If it is desired to "current bias" the element (drive a fixed DC current (I_{bias}) through the electrodes) a Keithley 2400 source meter is placed in parallel with the electrometer. If "voltage bias" (maintenance of a fixed potential (V_{bias}) across the electrodes) is desired then a Keithley 6517 electrometer is used to simultaneously establish V_{bias} and measure the DC current through the electrodes.

Results from four different types of test are reported here. In the first type of test, the furnace temperature is ramped at 120 °C/hr and the atmosphere is switched (at 5 min. intervals) between 0 and 450 ppm$_V$ NO$_x$ (either NO or NO$_2$, with 7 vol% O$_2$). Only the voltage (with no bias) is monitored during this test. The second type of test is isothermal and designed to investigate the effect of varying I_{bias}. The current source is programmed to step through discrete bias levels (e.g., -20, -10, -5, 0, +5, +10, +20 µA), dwelling for a fixed time at each. A brief pulse of NO$_x$ (again, either NO or NO$_2$, with 7 vol% O$_2$) is applied during the dwell at each bias level. This is done to identify which bias levels may be of further interest and when these are identified the third type of test is conducted: The operating temperature and I_{bias} are fixed, and the input NO or NO$_2$ concentrations are varied at fixed [O$_2$] or [O$_2$] varied at a fixed input [NO$_x$]. For the fourth and final type of test, the furnace is programmed to dwell at successively lower temperatures (e.g., 650, 630, 610, 600, 590, and 570 °C) and the input NO or NO$_2$ concentration is varied at each dwell temperature. During this test, designed to assess the effects of changing temperature, a constant V_{bias} is maintained and [O$_2$] is fixed at 7 vol%.

RESULTS AND DISCUSSION

Figure 3 shows the voltages measured when pulsing NO$_2$ during heating (Fig. 3a) and NO during cooling (Fig. 3b). The signals are difficult to interpret below ~550 °C in the case of input NO$_2$ and ~450 °C in the case of input NO but between these temperatures and about 650 °C NO$_2$ produces a small positive voltage and NO a smaller negative voltage. This would indicate that the element is behaving as a (rather poor) "non-Nernstian" or "mixed-potential" sensing element,[8-11] with NO and NO$_2$ responses of differing sign and the magnitude of the NO response less than that of the NO$_2$ response.[7]

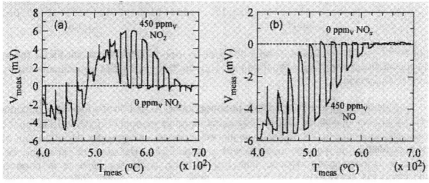

Figure 3: Measured voltages when pulsing NO$_x$ and ramping T at 120 °C/hr. Data in (a) taken while heating and in (b) taken while cooling. [O$_2$] = 7 vol%, N$_2$ balance.

Application of a DC bias current can change the relative sign and magnitudes of the NO and NO$_2$ responses from what is observed in Fig. 3. This is illustrated in Fig. 4, which shows measured traces ($V_{meas} = f(t)$) as 77 ppm$_V$ NO$_x$ is pulsed at different I_{bias} (Fig. 4a), and how the change in measured voltage due to the NO$_x$ (ΔV),

$$\Delta V = V_{n\ ppm_V\ NO_x} - V_{0\ ppm_V\ NO_x}, \qquad (1)$$

varies as a function of I_{bias} when $n = 77$ (Fig. 4b). For "negative biases" (the larger of the electrodes in Fig. 2a biased negatively with respect to the smaller), the changes in measured voltage induced by 77 ppm$_V$ NO$_2$ are larger than those induced by 77 ppm$_V$ NO. This is also true at sufficiently large positive biases. However, for a small range of positive bias currents (~1-4 μA) the introduction of 77 ppm$_V$ NO or NO$_2$ causes approximately the same change in the measured voltage. Therefore, it appears that the element is functioning as a "total NO$_x$" sensing element for [NO$_x$] = 77 ppm$_V$.

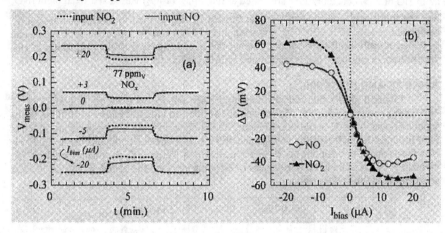

Figure 4: Measured voltage traces with 77 ppm$_V$ NO$_x$ at different current biases (a) and the computed changes in V_{meas} (ΔV, from Eqn. (1)) due to 77 ppm$_V$ NO$_x$ as a function of I_{bias} (b). T=600 °C, 7 vol% O$_2$.

 Figure 5 compares NO and NO$_2$ responses at 600 °C and 7 vol% O$_2$ with a current biased (+2.5 μA) sensing element over the concentration range ~1 ppm$_V$ $\leq$ [NO$_x$] $\leq$ 190 ppm$_V$. (The voltage changes due to NO$_x$ in Fig. 5 are plotted in percent, meaning that the ΔV of Eqn. (1) has been divided by the voltage measured with 0 ppm$_V$ NO$_x$.) Below about 120 ppm$_V$, the NO$_2$ response is consistently stronger than that for NO, while the opposite is observed for concentrations above this value.* Despite these systematic small differences in the NO$_x$ responses shown in Fig. 5, the fact that they are approximately equal is significant.
 The percent changes in measured voltage as a function of [NO$_x$] shown in Fig. 5 were well described by the equations

$$\Delta V = -(1 + a[NO_x])/(1 + b[NO_x]) \text{ (20–190 ppm}_V\text{, Fig. 5a), and} \qquad (2a)$$
$$\Delta V = -(c + a[NO_x])/(1 + b[NO_x]) \text{ (1–7 ppm}_V\text{, Fig. 5b).} \qquad (2b)$$

The fits shown in Fig. 5a used $a = 8.6 \times 10^{-1}$, $b = 1.1 \times 10^{-2}$ for NO and $a = 1.3$, $b = 2.1 \times 10^{-2}$ for NO$_2$ while those shown in Fig. 5b used $a = 5.0 \times 10^{-1}$, $b = 1.8 \times 10^{-2}$, $c = 4.4 \times 10^{-2}$ for NO and $a = 6.3 \times 10^{-1}$, $b = 2.7 \times 10^{-2}$, $c = 1.9 \times 10^{-2}$ for NO$_2$. The inability of Eqn. (2a) to fit the low NO$_x$ data of Fig. 5b) arises because below 4 ppm$_V$ the voltage changes induced by both NO and NO$_2$ vary in a linear fashion with [NO$_x$], and extrapolate much closer to 0 than -1 at [NO$_x$] = 0. Since Eqn. (2a) must pass through -1 at [NO$_x$] = 0, it diverges from the measured data.

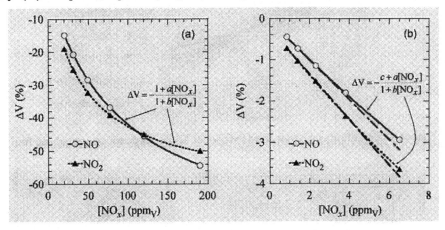

Figure 5: NO$_x$ response characteristics for a current-biased (+2.5 μA) sensing element operating at 600 °C in 7 vol% O$_2$. The straight lines drawn in b are an extrapolation from a linear fit to the three lowest concentrations.

The sensing mechanism giving rise to the total NO$_x$ behavior of these elements is presently unclear, but two mechanisms are under investigation. The first involves equilibration of the ratio [NO]/[NO$_2$] to a fixed value (irrespective of the input [NO]/[NO$_2$]) by the electrode materials. The second would suggest that NO$_2$ is reduced on one electrode and NO is oxidized on the other. The stronger NO$_2$ responses observed in $\Delta V = f(I_{bias})$ (Fig. 4b) suggests to us that the second of these hypotheses is more likely.

At 600 °C and fixed [NO$_x$], the [O$_2$] sensitivity of these sensing elements is a decreasing function of [NO$_x$]. This is illustrated in Fig. 6, which shows that in the absence of NO$_x$, varying [O$_2$] from 7 to 20 vol% produced changes in the measured voltage on the order of 30%. In the presence of 450 ppm$_V$ NO$_x$, similar variations in [O$_2$] produced changes only on the order of 10%. This behavior is different than that reported by Ho et al.[12] and Miura et al.[13] for Pt-containing, DC-biased sensing elements operating at 400 and 500 °C respectively and suggests that the incorporation (or liberation) of oxygen into (or from) the YSZ electrolyte is not the rate limiting step for current flow in these all-oxide sensing elements.

Figure 7a shows how the measured current (V_{bias} = 55 mV) varied with T at [NO$_x$] levels of 0, 20, 31, 77, and 190 ppm$_V$. When this data is cast in the form of an Arrhenius plot (Fig. 7b) it shows that the element DC resistance is an exponential function of T in the temperature range near 600 °C. Figure 7b also shows that the apparent activation energy of the electrical resistance is a decreasing function of [NO$_x$], varying from about 1.7 eV with 0 ppm$_V$ NO$_x$ to 1.0 eV with 190 ppm$_V$ NO$_x$.

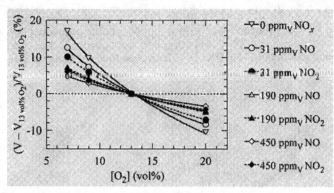

Figure 6: Effect of varying [O$_2$] at fixed [NO$_x$] (600 °C). Lines drawn are logarithmic fits.

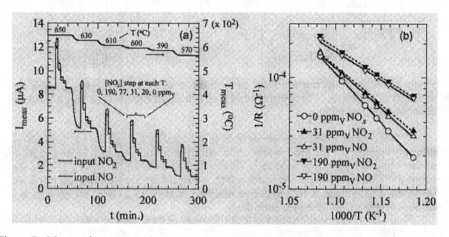

Figure 7: Measured currents and temperatures as [NO$_x$] is stepped at different temperatures (a). Arrhenius plots derived from the data in a are shown in b, with linear fits. Data collected with V_{bias} = 55 mV and 7 vol% O$_2$.

The DC electrical resistance of these sensing elements varies exponentially with T. The *changes* in element resistance due to NO$_x$ however are nearly linear with T, with higher [NO$_x$] yielding a more linear relationship (Fig. 8a). It is clear that temperature control will be important in the design of sensors incorporating these elements. Finally, Fig. 8b shows calculated element DC electrical resistances at 600 °C as a function of [NO$_x$] with both voltage and current biasing at 600 °C. The calculated resistances with both current and voltage biasing are nearly equal, indicating that these elements may be either voltage or current biased with equal effectiveness.

In view of the data in Fig. 8b, perhaps these sensing elements are most accurately are viewed as *resistive* elements, as opposed to amperometric elements if voltage biased or potentiometric elements if current biased.

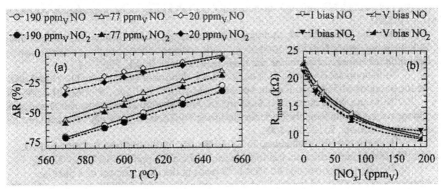

Figure 8: Changes in resistance induced by NO_x as a function of T (a) and measured resistance as a function of $[NO_x]$ at 600 °C with both voltage and current biasing (b).

CONCLUSION

All-oxide NO_x sensing elements operating at 600 °C have been developed. Use of DC electrical biasing (either constant current or constant voltage) enables these elements to exhibit "total NO_x" behavior, with responses to NO and NO_2 of the same algebraic sign and nearly the same magnitude. The elements respond to varying $[O_2]$ between 7 and 20 vol%, but the sensitivity is a decreasing function of $[NO_x]$. The DC electrical resistance of the elements varies exponentially with T and, over a narrow temperature range, was shown to exhibit an Arrhenius-like dependence on T. These elements could be useful in NO_x sensors for vehicle and other exhausts, and are currently under further study and development.

FOOTNOTES

*The stronger NO response persisted up to the highest concentration (1500 ppmv) measured, with this concentration of NO_x producing changes in V_{meas} of -80% for NO and -67% for NO_2.

ACKNOWLEDGEMENTS

B. L. Armstrong and C. A. Walls (Oak Ridge National Laboratory) contributed to the present work in the areas of substrate fabrication and ink development. Helpful discussions with D. Kubinski, R. Soltis, J. Visser, and B. Novak (Ford Scientific Research Laboratories) were also instrumental. Oak Ridge National Laboratory is operated by UT-Battelle, LLC for the United States Department of Energy under contract DE-AC05-00OR22725.

REFERENCES

[1]J. Kaspar, P. Fornasiero, and N. Hickey, "Automotive Catalytic Converters: Current Status and Some Perspectives," *Catalysis Today*, 77, 419-49, 2003.

[2]Calculations performed with FactSage, GTT-Technologies, TechnologiePark Herzogenrath, Germany.

[3]N. Kato, H. Kurachi, and Y. Hamada, "Thick Film ZrO_2 Sensor for the Measurement of Low NO_x Concentration," *SAE Tech. Pap. Ser.* 980170, 1998.

[4]N. F. Szabo and P. F. Dutta, "Strategies for Total NO_x Measurement with Minimal CO Interference Utilizing a Microporous Zeolitic Catalytic Filter," *Sens. Actuators B; Chem.*, **88**, 168-77, 2002.

[5]B. Nair, J. Nachlas, M. Middlemas, S. Bhavaraju, and C. Lewinsohn, "Mixed-Potential Type Ceramic Sensor for NO_x Monitoring," *28th International Cocoa Beach Conference and Exposition on Advanced Ceramics and Composites*, Cocoa Beach, FL, 2004.

[6]A. Kunimoto, M. Hasei, Y. Yan, Y. Gao, T. Ono, and Y. Nakanouchi, "New Total-NO_x Sensor Based on Mixed Potential for Automobiles," *SAE Tech. Pap. Ser.* 1999-01-1280, 1999.

[7]N. Miura, M. Nakatou, and S. Zhuiykov, "Impedancemetric Gas Sensor Based on Solid Electrolyte and Oxide Sensing Electrode for Detecting Total NO_x at High Temperature," *Sens. Actuators B; Chem.*, **93**, 221-8, 2003.

[8]F. H. Garzon, R. Mukundan, and E. L. Brosha, "Solid-State Mixed Potential Gas Sensors: Theory, Experiments, and Challenges," *Solid State Ionics*, **136-7**, 633-8, 2000.

[9]W. Gopel, R. Gotz, and M. Rosch, "Trends in the Development of Solid State Amperometric and Potentiometric High Temperature Sensors," *Solid State Ionics*, **136-137**, 519-31, 2000.

[10]N. Miura, G. Lu, and N. Yamazoe, "Progress in Mixed-Potential Type Devices Based on Solid Electrolyte for Sensing Redox Gases," *Solid State Ionics*, **136-7**, 533-42, 2000.

[11]E. Di Bartolomeo, M. L. Grilli, and E. Traversa, "Sensing Mechanism of Potentiometric Gas Sensors Based on Stabilized Zirconia with Oxide Electrodes," *J. Electrochem. Soc.*, **151**, H133-H9, 2004.

[12]K.-Y. Ho, M. Miyayama, and H. Yanagida, "NO_x Response Properties in DC Current of Nd_2CuO_4/4YSZ/Pt Element," *J. Ceram. Soc. Jpn.*, **104**, 995-9, 1996.

[13]N. Miura, G. Lu, M. Ono, and N. Yamazoe, "Selective Detection of NO by Using an Amperometric Sensor Based on Stabilized Zirconia and Oxide Electrode," *Solid State Ionics*, **117**, 283-90, 1999.

PHOTO-DEACTIVATED ROOM TEMPERATURE HYDROGEN GAS SENSITIVITY OF NANOCRYSTALLINE DOPED-TIN OXIDE SENSOR

Satyajit Shukla, Rajnikant Agrawal, Julian Duarte, Hyoung Cho, Sudipta Seal[*]
Mechanical Materials Aerospace Engineering (MMAE) Department and
Advanced Materials Processing and Analysis Center (AMPAC)
University of Central Florida (UCF)

Lawrence Ludwig
Kennedy Space Center (KSC-NASA)

[*]To whom the correspondence should be addressed

ABSTRACT

Nanocrystalline indium oxide (In_2O_3)-doped tin oxide (SnO_2) thin film is sol-gel dip-coated on the microelectromechanical systems (MEMS) devices as a room temperature hydrogen (H_2) sensor. The effect of ultraviolet (UV) radiation on the room temperature H_2 gas sensitivity of the present micro-sensor device is systematically studied. It is shown that the exposure to the UV-radiation results in the deterioration of the H_2 gas sensitivity of the present sensor, which is in contrast with the earlier reports. Very high H_2 gas sensitivity as high as 110×10^3 is observed, for 900 ppm H_2, without exposing the sensor to the UV-radiation. In the presence of UV-radiation, however, the H_2 gas sensitivity reduces to 200. The drastic reduction in the H_2 gas sensitivity due to the UV-exposure is explained on the basis of the constitutive equation for the gas sensitivity of the nanocrystalline semiconductor oxides thin film sensors.

INTRODUCTION

Nanocrystalline tin oxide (SnO_2) is a well known n-type semiconductor oxide, used in doped[1,2] and undoped[3,4] forms, for the gas sensing application. The mechanism of gas sensing using n-type semiconductor oxides, such as SnO_2, is now well established and involves decrease in the electrical resistance of the thin film due to the reaction of the reducing gas, such as hydrogen (H_2), with the surface-adsorbed oxygen-ions ($O_2^-{}_{ads}$ or $O^-{}_{ads}$ species), which releases the electrons into the conduction band.[5] The gas sensitivity (S) of the thin film, defined as the ratio R_{air}/R_{gas} or G_{gas}/G_{air} (where, R_{air} and R_{gas} are the resistances, and G_{air} and G_{gas} are the conductances of the thin film gas sensor in air without and with the reducing gas respectively), can be well predicted using the constitutive equation,[6-8] which is of the form,

$$S = A_1 \cdot \frac{2d}{D} \cdot \frac{C^n}{n_b} \cdot \exp\left[\frac{q^2}{2\varepsilon_r \varepsilon_0 k} \cdot \frac{\left[O^-\right]^2}{[V_0]T}\right] \qquad (1)$$

where, A_1 is a constant (m^{-3}), d the space-charge-layer thickness, D the nanocrystallite size, C the reducing gas concentration (ppm), n the gas concentration exponent, n_b the bulk charge-carrier-concentration, q the electronic charge, $\varepsilon_r \varepsilon_0$ the permittivity of the sensor, k the Boltzmann's constant, $[O^-]$ the surface-density of states, $[V_0]$ the oxygen-ion vacancy concentration, and T the sensing temperature.

On the other hand, there is an increased demand for H_2 (in liquid and gaseous forms) as a cheap replacement fuel for automobiles, power generation using solid oxide fuel cells (SOFCs), and launching the space-shuttles into the space. As a result, much attention has been recently given for the production, storage, and transportation of H_2.[9-11] Since, H_2 is the smallest gaseous molecule, it is more susceptible for leakage through the existing system of pipelines and storage devices. As H_2 can catch fire easily in the presence of oxygen when present in critical amount (4 %), detecting H_2 leakage is vital for safety concerns. Since H_2 gets combusted easily at higher temperatures, sensing H_2 at lower temperatures (typically room temperature) is essential. However, H_2 sensors based on nanocrystalline-SnO_2 exhibits higher H_2 sensitivity at relatively higher operating temperatures (250-300 °C).[12] The room temperature H_2 detection with high sensitivity using nanocrystalline-SnO_2 sensor is still lacking in the literature. From this point of view, the major objective of the present article has been set to demonstrate a giant room temperature H_2 sensitivity of the nanocrystalline-SnO_2 based sensor. Although in the literature[13-15], the exposure to the UV-radiation has been shown to enhance the room temperature gas sensitivity of the SnO_2-based sensors, in this article we demonstrate that, such UV-exposure deteriorates the room temperature H_2 sensitivity of the present sensor. The observed sensor behavior has been explained on the basis of the constitutive equation for the gas sensitivity of the nanocrystalline semiconductor oxides thin film sensors.

EXPERIMENTAL

A 3" Si (100) wafer was base-cleaned and used as a substrate for microelectromechanical system (MEMS) device fabrication. SiO_2 was thermally grown as an insulation layer. 20 nm chromium (Cr) / 200 nm gold (Au) films were deposited by thermal evaporation on the top of SiO_2 layer. The interdigitated Au-electrodes, Figure 1, were patterned with the electrode distance of 20 µm on the metallic layers using photolithography and wet chemical etching.

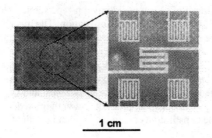

1 cm

Figure 1. Silhouette of the MEMS device, which is sol-gel dip-coated with the nanocrystalline In_2O_3-doped SnO_2 thin film sensor. The MEMS device has four interdigitated electrodes and one resistive temperature sensor in the center.

In order to deposit a nanocrystalline thin film of indium oxide (In_2O_3)-doped SnO_2, tin(IV)-isopropoxide ($Sn[OC_3H_7]_4$) (10 w/v %) in iso-propanol (72 vol. %) and toluene (18 vol. %) and indium(III)-isopropoxide ($Sn[OC_3H_7]_3$) were used as starting materials. 0.23 M solution of tin-isopropoxide in iso-propanol and toluene mixture was first prepared. Appropriate amount of indium(III)-isopropoxide was added to this solution to obtain the thin films containing 6.5 mole % In_2O_3. Sol-gel dip coating technique was adopted with a pulling speed of 150 cm/min for thin film deposition. The films were dried at 150 °C for 1 h in air. The dip-coating was repeated until the desired film thickness was achieved. Over the dried gel film, a thin layer of Pt was sputter-coated for 10 seconds. Finally, the Pt-sputtered dried gel films were heat treated at 400 °C in air. The coated MEMS-devices were heated at a rate of 30 °C/min up to the firing temperature, held at that temperature for 1 h, and then, cooled down to room temperature inside the furnace. After sol-gel dip-coating of In_2O_3-doped SnO_2 thin films over the sensor platform, the thin films were characterized using x-ray photoelectron spectroscopy (XPS), atomic force microscopy (AFM), focused ion-beam (FIB) milling, and transmission electron microscopy (TEM) techniques. The coated-MEMS device was wire-bonded to a ceramic package for final H_2 sensing tests.

The micro-sensor device was tested for sensing 900 ppm H_2 at room temperature (22 °C) in 50 Torr air under dynamic test-condition. The original sensor-resistance stabilized in air was first measured. Then, 900 ppm of H_2 was admitted into the test-chamber under the dynamic test-condition. After the total response time of 60 min, the H_2 flow was stopped and air at 760 Torr was blown in to retain the original sensor-resistance in air. The chamber pressure was then reduced again to 50 Torr and the sensor-test was repeated for three-cycles. The above sensor-test was conducted without and with the UV-exposure. For the stimulation of the sensor-surface with the UV-radiation, an InGaN based light emitting diode (LED) mounted on the lead frame with clear epoxy lens was used. On forward bias, it emitted a band of visible light that peaked 375 nm.

RESULTS AND DISCUSSION
     A typical TEM-image of FIB-milled TEM-sample, showing the cross-section of the sol-gel dip-coated MEMS device is presented in Figure 2.

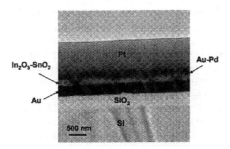

Figure 2. A typical TEM image of FIB-milled TEM-sample, showing the cross-section of the sol-gel dip-coated MEMS device. In_2O_3-doped SnO_2 thin film sensor is deposited over the Au-electrode patterned over Si/SiO_2 substrate. Au-Pd and Pt are the protective layers for FIB-milling.

The In_2O_3-doped SnO_2 thin film sensor is observed to be deposited over the Au-electrode and has an average thickness of 125-150 nm. At a given magnification, the thin film appears to uniform and continuous. Typical AFM image of the In_2O_3-doped SnO_2 thin film is shown in Figure 3. An average In_2O_3-doped SnO_2 nanoparticle size of 50 nm is noted in Figure 3.

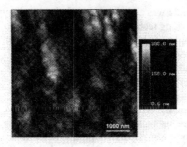

Figure 3. AFM image of In_2O_3-doped SnO_2 thin film sensor sol-gel dip-coated on the MEMS device. The average nanoparticle size and the film thickness are observed to be 50 nm and 125-150 nm respectively.

HRTEM analysis, as reported elsewhere[16], revealed that, these 50 nm particles are made up of 6-8 nm nanocrystallites. Further, the average film thickness as measured via AFM analysis, Figure 3, is in good agreement with the TEM results. Thus, the film thickness and the nanocrystallite size are optimum for sensing H_2 at room temperature with high sensitivity.[12]

The broad scan XPS analysis[17] of the nanocrystalline In_2O_3-doped SnO_2 thin film sensor, sol-gel dip-coated on the MEMS device, revealed the presence of Sn, In, O, and Pt as major-peaks on the sensor-surface, which have been originated from the sol-gel dip-coating process. There was no evidence of the presence of Si and Au, which may originate from the substrate (MEMS Device). Hence, the sol-gel deposited thin film appears to be continuous and thicker (relative to x-rays) covering the entire Si/SiO_2 substrate. The narrow scan XPS analysis of Sn (3d), In (3d), and Pt (3d) peaks were conducted within the B.E. range of 480-500 eV, 435-460 eV, and 65-85 eV respectively. Sn $3d_{5/2}$ and In $3d_{5/2}$ B.E. levels of 487.1 eV and 445 eV were observed, which correspond to Sn^{4+} and In^{3+} ionic-states respectively.

The present sensor was first exposed to the UV-radiation with alternate ON and OFF states in order to confirm its response to the UV-exposure. Under the UV-radiation, the sensor-resistance is observed to decrease due to the transition of the valence band electrons into the conduction band and desorption of some of the surface-adsorbed oxygen-ions; while with the UV-OFF state, the sensor-resistance increased back to its original value. The cyclic sensor-tests, for detecting H_2 at room temperature, were then conducted.

Typical results of a cyclic sensor-test, conducted at 22 °C and 50 Torr air pressure for 900 ppm H_2 without the UV-exposure is presented in Figure 4. The total response time is kept constant to 60 min for the three-cycles. Four- to five-orders of magnitude drop in the sensor-resistance is noted in the presence of 900 ppm H_2. A very high value of room temperature H_2 sensitivity as high as 110×10^3 is observed in this investigation, which according to us, is much higher than those reported under any processing and test conditions for the SnO_2-based sensors.

When exposed to the UV-radiation, the sensor-resistance in air is observed to decrease from 1 GΩ to 1.5 MΩ, Figure 5. The sensor-resistance in air decreases first abruptly upon the exposure to the UV-radiation, which has been attributed to the transition of the valence band electrons to the conduction band. This is followed by a gradual decrease in the sensor-resistance

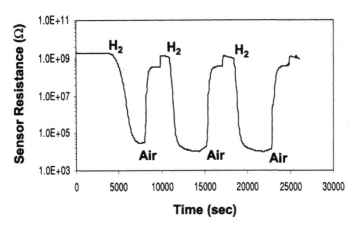

Figure 4. Typical cyclic-response of the present micro-sensor device to 900 ppm H_2, without the UV-exposure, under the dynamic test-condition. The sensor-test is conducted at 22 °C and 50 Torr air pressure.

over the period of time, which is attributed to the continuous desorption of the surface-adsorbed oxygen-ions under the UV-exposure.

Typical results of a cyclic sensor-test, conducted at 22 °C and 50 Torr air pressure for 900 ppm H_2 under the UV-exposure is presented in Figure 6. The total response time is again kept constant to 60 min for the three-cycles. Two-orders of magnitude drop in the sensor-resistance is noted in the presence of 900 ppm H_2 under the UV-exposure. The room temperature H_2 sensitivity under the UV-exposure is calculated to be 200, which is very low compared with that observed in Figure 4.

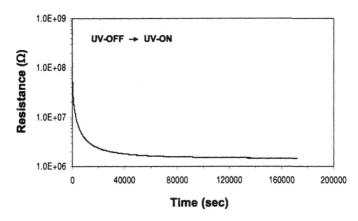

Figure 5. Variation in the original sensor-resistance in air after exposure to the UV-radiation.

61

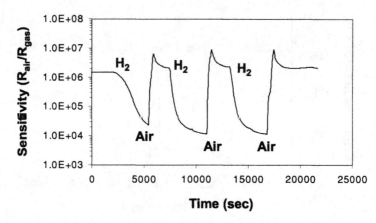

Figure 6. Typical cyclic-response of the present micro-sensor device to 900 ppm H_2, with the UV-exposure, under the dynamic test-condition. The sensor-test is conducted at 22 °C and 50 Torr air pressure.

The present micro-sensor device, thus, exhibits a photo-deactivated room temperature H_2 sensitivity under the UV-exposure. This has been primarily attributed the decrease in the original sensor-resistance upon exposure to the UV-radiation (Figure 5), which is a result of desorption of the surface-adsorbed oxygen-ions. The reduced room temperature H_2 sensitivity due to the loss in the surface-adsorbed oxygen-ions can be very well explained on the basis of the constitutive equation for the gas sensitivity of the semiconductor oxides thin film sensors, Equation 1. It is to be noted that, although the total H_2 exposure time is 60 min, the response and the recovery time of 5-15 min are recorded under both the conditions at room temperature for the H_2 sensitivity levels suitable for the practical applications.

We attribute the room temperature H_2 sensitivity of the present MEMS-based sensor to the effect of the combination of number of effective variables involved such as the UV-exposure, the small electrode distance provided by the MEMS device, the nanocrystalline nature of the sensor material, and the presence Pt-surface catalyst. The exposure of the sensor-surface to the UV-radiation is known to burn the organic residue from the sensor-surface; thus, providing a clean sensor-surface for detecting H_2 at the lower operating temperatures.[18,19] The small electrode distance of 20 μm is conducive to enhance the transducer function of the sensor.[20] On the other hand, the nanocrystallite size of less than 10 nm has been shown to drastically increase the gas sensitivity of the SnO_2-based sensors.[5,21] Lastly, the surface Pt-clusters are responsible for reducing the activation energy of the dissociation of H_2 leading to improved response kinetics at room temperature.[5] The overall surface reactions involved in the H_2 sensing using the present micro-sensor device can be summarized as:

$$H_{2\,(gas)} + O^-_{\,(ads)} \rightarrow H_2O_{\,(ads)} + e^- \quad \text{(Response)} \qquad (2)$$

$$O_{2\,(gas)} + e^- \rightarrow 2O^-_{\,(ads)} \quad \text{(Recovery)} \qquad (3)$$

CONCLUSIONS
    It is concluded that, the present nanocrystalline In_2O_3-doped SnO_2 thin film based micro-sensor device is a good room temperature H_2 sensor, with a very high room temperature H_2 sensitivity of 110×10^3 with the response and the recovery time in few minutes at room temperature. The room temperature H_2 sensitivity of the present MEMS-based sensor has been attributed to the effect of the combination of number of effective variables involved such as the UV-exposure, the small electrode distance provided by the MEMS device, the nanocrystalline nature of the sensor material, and the presence Pt-surface catalyst. However, very low room temperature H_2 sensitivity (200) has been observed for the present micro-sensor device under the UV-radiation due to the reduced concentration of the surface-adsorbed oxygen-ions, which is in agreement with the prediction of the constitutive equation for the gas sensitivity of the nanocrystalline semiconductor oxides thin film sensors.

ACKNOWLEDGEMENTS
    Authors thank NASA-Glenn, National Science foundation (Grant # NSF CTS 0350572), and Florida Space Grant Consortium (FSGC) for the financial support.

REFERENCES
    [1]S. Shukla, S. Seal, L. Ludwig, and C. Parrish, "Room Temperature Hydrogen Gas Sensor Based on Nanocrystalline 6.5 mol % Indium Oxide Doped-Tin Oxide Thin Film", *Sens. Actuators B* **97**, 256-265 (2004).
    [2]K.H. Cha, H.C. Park, K.H. Kim, "Effect of Palladium Doping and Film Thickness on the H_2-Gas Sensing Characteristics of SnO_2", *Sens. Actuators B* **21**, 91-96 (1994).
    [3]G. Sakai, N.S. Baik, N. Miura, N. Yamazoe, "Gas Sensing Properties of Tin Oxide Thin Films Fabricated From Hydrothermally Treated Nanoparticles Dependence of CO and H_2 Response on Film Thickness", *Sens. Actuators B* **77**, 116-121 (2001).
    [4]J.-P. Ahn, S.-H. Kim, J.-K. Park, M.-Y. Huh, "Effect of Orthorhombic Phase on Hydrogen Gas Sensing Property of Thick-Film Sensor Fabricated by Nanophase Tin Oxide", *Sens. Actuators B* **94**, 125-131 (2003).
    [5]S.Shukla and S. Seal, "Nanocrystalline SnO Gas Sensor in View of Surface Reactions and Modifications", *JOM* **54**, 35-38, 60 (2002)
    [6]S. Shukla and S. Seal, "Theoretical Model for Nanocrystallite Size Dependent Gas Sensitivity Enhancement in Nanocrystalline Tin Oxide Sensor", *Sens. Letts.* **2**, 73-77 (2004).
    [7]7. S. Shukla and S. Seal, "Constitutive Equation for Gas Sensitivity of Nanocrystalline Tin Oxide Sensor", *Sens. Letts.* **2**, 125-130 (2004).
    [8]S. Shukla and S. Seal, *Sens. Letts.* "Theoretical Model for Film Thickness Dependent Gas Sensitivity Variation in Nanocrystalline Tin Oxide Sensor", *Sens. Letts.* **2**, 260-264 (2004).
    [9]E.D. Wachsman and M.C. Williams, "Hydrogen Production From Fossil Fuels With High Temperature Ion Conducting Ceramics", *Interface* **13**, 32-37 (2004).
    [10]L. Schlapbach, "Hydrogen as a Fuel and Its Storage for Mobility and Transport", *MRS Bulletin* **27**, 675-680 (2002)
    [11]R.S, Irani, "Hydrogen Storage: High-Pressure Gas Containment", *MRS Bulletin* **27**, 680-684 (2002)
    [12]S. Shukla and S. Seal, "Nanocrystalline Semiconductor Tin Oxide as Hydrogen Sensor", in *Encyclopedia of Sensors*, edited by C. Grimes, E. Dickey, and M. Pishko, (American Scientific Publisher, Stevenson Ranch, CA, 2005) accepted.

[13]E. Comini, G. Faglia, and G. Sberveglieri, "UV Light Activation of Tin Oxide Thin Films for NO_2 Sensing at Low Temperatures", *Sens. Actuators B* **78**, 73-77 (2001).

[14]P. Camagni, G. Faglia, P. Galinetto, C. Perego, G. Samoggia, and G. Sberveglieri, "Photoensitivity Activation of SnO_2 Thin Film Gas Sensors at Room Temperature", *Sens. Actuators B* **31**, 99-103 (1996).

[15]E. Comini, A. Cristalli, G. Faglia, and G. Sberveglieri, "Light Enhanced Gas Sensing Properties of Indium Oxide and Tin Oxide Sensors", *Sens. Actuators B* **65**, 260-263 (2000).

[16]S. Shukla, S. Patil, S. Kulry, S. Seal, L. Ludwig, and C. Parrish, "Synthesis and Characterization of Sol-Gel Derived Nanocrystalline Tin Oxide Thin Film as a Hydrogen Gas Sensor", *Sens. Actuators B* **76**, 343-353 (2003).

[17]S. Shukla, R. Agrawal, L. Ludwig, H. Cho, and S. Seal, "Effect of Ultraviolet Radiation on Room Temperature Hydrogen Sensitivity of Nanocrystalline Sol-Gel-Doped Tin Oxide MEMS Sensor", *J. Appl. Phys.* (in press).

[18]G.K. Mor, M.A. Carvalho, O.K. Varghese, M.V. Pishko, and C.A. Grimes, "A Room-Temperature TiO_2-Nanotube Hydrogen Sensor Able to Self-Clean Photoactively from Environmental Contamination", *J. Mater. Res.* **19**, 628-634 (2004).

[19]O.K. Varghese, D. Gong, M. Paulose, K.G. Ong, C.A. Grimes, "Hydrogen Sensing using Titania Nanotubes", *Sens. Actuators B* **93**, 338-344 (2003).

[20]X. Vilanova, E. Llobet, J. Brezmes, J. Calderer, and X. Correig, "Numerical Simulation of the Electrode Geometry and Position Effects on Semiconductor Gas Sensor Response", *Sens. Actuators B* **48**, 425-431 (2000).

[21]F. Lu, Y. Liu, M. Dong, and X. Wang, "Nanosized Tin Oxide as the Novel Material with Simultaneous Detection towards CO, H_2 and CH_4", *Sens. Actuators B* **66**, 225-227 (1998).

PTCR-CO CERAMICS AS CHEMICAL SENSORS

Zhi-Gang Zhou[*]
State Key Laboratory of Fine Ceramics and New Processing
Department of Materials Science and Engineering
Tsinghua University
Haidian District
Beijing 100084, China

Zi-Long Tang
State Key Laboratory of Fine Ceramics and New Processing
Department of Materials Science and Engineering
Tsinghua University
Haidian District
Beijing 100084, China

Zhong-Tai Zhang
State Key Laboratory of Fine Ceramics and New Processing
Department of Materials Science and Engineering
Tsinghua University
Haidian District
Beijing 100084, China

ABSTRACT

A new CO gas sensor of Perovskite-type oxide, semiconducting doped $BaTiO_3$ based PTCR ceramics was studied. PTCR ceramic is capable of detecting carbon monoxide in the concentration range of 1-5% in the PTCR / NTCR region below/above the critical temperature, by means of higher barrier potential of PTCR / NTCR effect. The phenomenon is based on anionic adsorption, the extreme activity of the oxygen atoms at the surfaces of grain boundaries at the lower temperature while the bulk defect structures are frozen-in. The interaction between CO gases with the oxygen at the surface layer of the PTCR ceramics participates in the sensing reaction, and the itinerant electron comes from the conduction band of the n-type semiconducting ceramics. The creation of the new electron traps in the grain boundaries bring about to a decrease in resistivity of the material. The physisorption between the CO gas on the surface of the PTCR grains is the key for the ceramics working at lower temperature. The adsorption of CO gas on PTCR ceramics and mechanism of PTCR-CO working in lower temperature as well as kinetic process of defects are discussed. [Key words: positive temperature coefficient resistance, barium titanate, CO gas sensor, defects, adsorption.]

I. INTRODUCTION

PTCR ceramics are widely used in industries and civilian as a control device. Now the PTCR ceramics have developed to be a new multifunctional semiconducting ferroelectric ceramics. That

---

* Member, American Ceramic Society

---

is say; the PTCR ceramic not is a physical sensor only since 50's last century, but for a chemical sensor just in recent years by a limited number of investigators [1-2].

As is well know that, CO gas is a deadly colorless odorless and one of the important toxic gases from the exhaust. It is the most interesting caused by threat of global warming and permanent changes in our environment loom overhead and damage the human body. So, a few of metal oxide semiconductors as n-type SnO_2 [3], ZnO [4], TiO_2 [5], and p-type as (Ln M) BO3 [6] have been studied into a long time for detecting CO gas.

Recently, for CO sensor development, a few researchers now are working on decrease working temperature of SnO_2 based materials by means of promotion of the activity of CO reaction on the surface [7-8] as well as the sensitivity, it is interested to increase the porosity of the ceramics and adding a small amount of metal catalysis (such as Pt, Pd, Ag, Au, Er and Cu).[6-10] And now a few of them are working on low price CO gas sensor using by new application of doped $BaTiO_3$ semiconductor with posistor effect [11-12].

For the mentioned above, in this paper, the La doped $BaTiO_3$ ceramics with PTCR effect was used for CO detection. The nature is revealed on CO gas response of PTCR ceramics in the low/middle temperature. The adsorption of CO gas on PTCR ceramics and the phenomenon of PTCR-CO working in low/middle temperature as well as kinetic process of defects are studied.

II. EXPERIMENTAL

The n-type semiconducting $BaTiO_3$ PTCR ceramics such as $Ba_{0.92} La_{0.08} TiO_3$ was prepared by conventional ceramic semiconducting technology. The resistivity measurements were conducted on a ceramic pellet with an In-Ga alloy rubbed on both end surfaces as electrodes. This two-probe method was used as in a dc bias of 10 V in varied concentration of CO gas in air, of course, air was used as a carrier gas with flow rate of 30 ml min^{-1} during working. The measurements were performed on all samples, in the tubular flow quartz-glass chamber with thermocouple, first in air and then successively under the reducing CO gas.

The measurements were performed at 90-150°C in the 1 and 5.0 vol. % concentration of CO gas. The temperature was monitored and the resistance was measured automatically every 30 sec. The samples (10 ϕ x 1mm) with porosity of about 15% sintered at 1150 °C for 1h. Impedance (Z) measurements were performed using an HP 4194 impedance analyzer at a signal strength of 0.5 V_{rms} under various kinds of ambient atmosphere, such as air, 1 and 5 vol.% CO (N_2 as carrier gas) at different temperatures. The operating frequency ranged from 100Hz to 40MHz. Preparation of starting material and the methods of characterization were the same as in our previous publications[1-2].

III. RESULTS AND DISCUSSION

(1) Impedance plots

Impedance plots of the sample in different atmosphere and different measuring temperatures are as shown in Fig.1 and 2. As is now known that the non-uniform relaxations are manifested by the skewed arcs during the charge transport within doped $BaTiO_3$ in reducing atmosphere. The area of the skewed arc is found to increase with increasing test temperature T. That is to say, while PTCR-CO ceramic working in PTCR region (T ρ_{max} >T > T_c, T ρ_{max} is the temperature of the maximum resistivity, and the Currie point T_c is about 120°C) the resistance of grain boundary R_{gb} increases continuously with increase of temperature but decreases with increase of temperature in NTCR region while test temperature T high than T^{ρ}_{max} ($T > T^{\rho}_{max} > T_c$) as shown in Figure 3.

(2) Response characteristics and energies

The response characteristics of resistivity to CO gas are another important property to be examined for this PTCR ceramics being considered as a CO gas sensor. From this viewpoint, responses to reducing ambient gas containing 5.0 vol. % concentration of CO gas have been obtained for sample at 90-150°C as shown in figure 4. The change of resistance slope in grain boundary(GB) is an indirect evidence for that PTCR ceramics could be activated above the critical temperature (above 120°C) whereas the lower slope means lower resistance at grain boundary after working in reduced gas as 5 % CO gas. And the grain boundary of the PTCR ceramics still play a main role for sensing process after expose in CO gas.

Further indirect information about the energies in the grain boundary phases can be obtained from the temperature dependence of the quantities R_g and R_{gb}. In crystalline extrinsic ionic conductors the conductivity is thermally activated as:

$$\sigma = A_e t / R = (\sigma_0 / T) \exp(-\Delta H_m / k_B T) \tag{1}$$

where H_m is the activation enthalpy for ionic migration. The Arrhenius plots of the ionic conductivity or resistivity linearize this equation and their slope gives $\Delta H_m / k_B$. Arrhenius plots of R_g and R_{gb} for both ceramics are linear over the temperature range examined (120-150°C) as shown in Figure 5. The lower activation energy of grain boundaries in CO gas, $E_{CO\text{-}gas}$ (0.73 eV) < E_{air} (0.93 eV), means the possibility of this sensing process for the interaction CO gas on the surfaces of the PTCR ceramics.

(3) Mechanism and role of defects

The mechanism of reactions with oxygen vacancies near the surface may be complicated, especially if bulk defect structures are often "frozen-in" at low - temperatures. The ceramics can therefore be designed that oxygen molecular makes use of physisorption, there is a weak interaction between the adsorbed molecules (such as CO gas) on the surface of the grains, working at lower temperatures. And the chemisorption, there is a strong interaction between the adsorbated molecules at available temperatures, which higher than the temperature of physisorption after overcoming a potential barrier of activation energy. It was recognized by thermal desorption spectroscopy (TDS) experiments. [13-18] The macro phenomenon at low temperatures is assumed to involve trapping of electrons after removal of oxygen atoms physisorbed on the grain surfaces by their reactions with CO molecules (Fig. 6). The results presented above can be summarized as follows with respect to the possible mechanism for CO detection by BaTiO3 based PTCR ceramics [19-23].

(i) For CO physisorption on the oxide surface:

$$CO_g \Rightarrow CO_{ads} \Rightarrow CO^+_{ads} + e^- \tag{2}$$

$$CO + 1/2\ O_2 \Rightarrow CO_2 \tag{3}$$

$$CO_{ads} + O_2^- \Rightarrow CO_2 + e^- \tag{4}$$

$$O_{ads}^- + CO \Rightarrow CO_2 + e^- \tag{5}$$

and

$$O_2 + e^- \Rightarrow O_{ads}^- \tag{6}$$

where CO_{ads}, CO_{ads}^+, O_{ads}^-, and O_2^- are carbon monoxide, monovalent carbon monoxide, monovalent oxygen at adsorbed states, and monovalent oxygen molecular on the surface of ceramics, respectively.

(ii) Redox reaction of CO on the oxide surface by chemisorption:

$$CO_g + O_O^x \Rightarrow CO_2 + e^- + V_O^+ \tag{7}$$

$$CO_g + 1/2\,O_2 + V_O^x + e^- \Rightarrow CO_2 + O_O^x \tag{8}$$

$$CO_g + O^{2-} + V_O^{2+} \Rightarrow CO_2^- + V_O^+ \tag{9}$$

$$CO + 2O^{2-} + 2V_O^{2+} \Rightarrow CO_3^{2-} + 2V_O^+ \tag{10}$$

where V_O^+, V_O^{2+}, V_O^x and O_O^x are monovalent oxygen vacancy, divalent oxygen vacancy, neutral oxygen vacancy, and lattice oxygen in the ceramic lattice, respectively. And CO_2^- is a carboxylate and CO_3^{2-} is a bidentate carbonate. [14-15]

(iii) In presence of molecular oxygen and forming with a surface vacancy:

$$O_2 + V_O^+ \Rightarrow O_{2\,ads}^- + V_O^{2+} \tag{11}$$

$$O_2 + 2V_O^+ \Rightarrow 2O_{ads}^- + 2V_O^{2+} \tag{12}$$

$$1/2\,O_2 + [OCOV_o]^+ + e^- \Rightarrow CO_2 + O_O^x \tag{13}$$

$$CO_g \Rightarrow [OCOV_o]^+ + e^- \tag{14}$$

where $O_{2\,ads}^-$ is monovalent molecular oxygen at adsorbed states. $[OCOV_o]^+$ is a clustered pair consisting of one $CO_{2\,g}$ bound to a surface oxygen vacancy V_o, which together are monovalent. [13] This type of defect complex is sometimes called a "dimer". [24]

Eqs. (7) - (14) include the lattice oxygen reaction and surface complex, which occur at available, not lower, temperatures after overcoming a potential barrier of activation energy. And the following reaction chain can, for instance, the state of adsorptions, can be changed from physisorption into chemisorption with increasing of temperature and/or CO/ O_2 partial pressure in the reactant mixture:

$$O_2 (g) \xrightarrow{(1e^-)} O_2^- (ads) \xrightarrow{(1e^-)} 2O^- (ads) \xrightarrow{(2e^-)} 2O^{2-} (ads) \qquad (15)$$
$$\text{(>150°C)} \quad \text{(>150°C)} \quad \text{(>400°C)}$$

According to the above reaction, it is apparent that the reducing CO gas leads to desorption of the oxygen on the grain surfaces and to a decrease in the concentration of the adsorbed oxygen, resulting in a decrease in the resistivity of the PTCR ceramics. As the amount of reaction products, CO_2 +e ‾, resulting from reactions (2) - (6) increases, the decrease in resistivity becomes remarkable, free electrons as a driving force for this kind of CO gas sensor, if in the lower temperature by physisorption only.[15-20] It should be noted that presence of O_{ads}^- and/or O^{2-} centers would be leaded to forming shallow trapped hole centers in $BaTiO_3$ and increase the concentration of electrons at lower temperature[25-26]. A possible mechanism of oxidation of CO physisorbed on the surfaces of the grains and the possible role of defects are as shown in Fig. 6, and 7, respectively.

That is, at low temperatures, chemisorption may be so slow that for practical purposes only physisorption is observed at high temperatures, and physisorption is small and chemisorption occurs only coursed by the lower adsorption energy. So, the PTCR ceramics can therefore be designed that oxygen molecular makes use of physisorption on the surface of the grains at lower temperatures and than chemisorption at available temperatures after overcoming a potential barrier of activation energy. We can conclude that the chemisorption may be slow and the rate behavior indicative of the presence of an activation energy; in other words, it is possible for physically adsorbed at first, and then, more slowly, to enter into some chemical reaction with the surface of the ceramics. A few valuable research reports recognized that perovskite ceramics as a good CO gas sensor could be operated in low temperature.[6, 19, 21]

In spite of the above discussion the mechanism of PTCR-CO response is not fully clarified yet, especially in lower temperature. However, the above results will stimulate further investigations, hopefully leading to PTCR, $BaTiO_3$ based or others, ceramics for CO to common usage.

IV. CONCLUSIONS

The perovskite, n-type semiconducting PTCR-CO porous ceramics, is a new perspective CO gas sensing material in the PTCR region with lower working temperature and cheaper cost. PTCR ceramic is capable of detecting carbon monoxide in the concentration range of 1-5% in the PTCR region below the maximum temperature of resistivity, by means of higher barrier potential of PTCR effect. Physisorption is a possible mechanism for oxidation of CO gas on the surfaces of the PTCR grains at lower temperature. More generally, much work remains to be done in improving the properties of the sensor in the application level, such as increase the sensitivity, selectivity, and stability and decrease the operating temperature in room temperature etc. For promoting the sensitivity, it is required to increase the porosity of the ceramics and adding a small amount of metal catalysis and nano-dots as new adsorption centers (such as Pt, Pd, Ag and Cu) are needed. And now the emergence of this new member of PTCR ceramics as a new CO gas sensor will push the PTCR ceramics from the traditional field to a new boundary and faces on a new challenge.

Acknowledgements: We are grateful to the financial support of the National Natural Science Foundation of China (NSFC) (Contract No. 59672012). We also thank many colleagues of the State Key Laboratory of New Ceramics and Fine Processing, Department of Materials science and Engineering of the Tsinghua University for their help.

REFERENCES

[1] Z.-G. Zhou, Z -L. Tang, Z,-T. Zhang, W. Wlodarski, "Perovskite Oxide of PTCR ceramics as Chemical Sensors", *Sensors and Actuators B, 11*, 22-26 (2001).

[2] Zhi-Gang Zhou, Zi-Long Tang, Zhong-Tai Zhang, "Studies on Grain-Boundary Chemistry of Perovskite Ceramics as CO Gas Sensors", *Sensors and Actuators B*, 93, 356-361(2003).

[3] N. Yamazoe, Y. Kurokawa, and T. Seiyama, "Effects of Additives on Semiconductor Gas Sensors", *Sensors and actuators*, 4, 283-289 (1983).

[4] Y. Nakamura, H. -X. Zhang, A. Kishimoto, O. Okada, H. Yamagida, "Enhanced CO and CO_2 Gas Sensitivity of CuO/ZnO Heterocontact made by Quenched CuO ceramics", *Journal of the Electrochemical Society*, 145, 632- (1998).

[5] V. Guidi, M.C. Caeota, M. Ferroni, G. Martinelli, L. Paglialoga, E. Comini, P. Nelli, G. Sberveglieri, *Nanosized materials for gas sensing applications*, in *Sensors and Microsystems*, Edited by S. D. Natale, A. D'Amico, G. Sberveglieri, World Scientific, p.135,1999.

[6] C. M. Chiu, Y.H. Chang, "The Structure, Electrical and Sensing Properties for CO of the $La_{0.8}Sr_{0.2}Co_{1-x}Ni_xO_{3-\delta}$ System", *Materials Science and Engineering, A*, 266, 93- 98 (1999).

[7] M. Haruta, T. Subota, T. Kobayashi, H. Kageyama, M. J. Genet and B. Delmon, "Low Temperature Oxidation of CO over Gold Supported on TiO_2, α-Fe_2O_3 and Co_3O_4", *J. Catalysis*, 144, 175-192 (1993).

[8] F. Boccuzzi, A. Chiorino, S. Tsubota, M. Haruta, "An IR Study of CO-sensing Mechanism on Au/ZnO", *Sensors and Actuators B*, 24-25, 540-543 (1995).

[9] N. Yamazoe, "New Approaches for Improving Semiconductor Gas Sensors", *Sensors and Actuators B*, 5, 7-19 (1991).

[10] T. Kabayashi, M. Haruta, H. Sano and M. Nakane, "A Selective CO Sensor Using Ti-doped α-Fe_2O_3 with Coprecipitated Ultra fine Particle of Gold" *Sensors and Actuators*, 13, 339-349(1988).

[11] Z.-G. ZHOU, Z.-L. Tang, Z.-T. Zhang , "PTCR-CO gas sensor", special issue of *Asian Journal of Physics* on " Chemical and Biosensors", 14, 1/2, 47-53 (2005).

[12] M. Viviani, M. T. Buscaglia, V. buscaglia, L. Mitoseriu, A. Testino, P. Nanmi, D. Vladikova, " Analysis of Conductivity and PTCR effect in Er-doped $BaTiO_3$ Ceramics, "*Journal of the European Ceramic Society*, 24, 1221-1225 (2004)

[13] W. Gopel, "Chemisorption and Charge Transfer at Ionic Semiconductor Surfaces: Implications in Designing Gas Sensors" *Progress in Surface Science*, 20[1], 9-103, (1985).

[14] M. J. Willett, "Spectroscopy of Surface Reactions" in Techniques and Mechanism in Gas Sensing, Edited by P. T. Mosely, J. O. W. Norris, D. E. William, Adam Hilger, Bristol, USA, p.61-107,1991.

[15] P. T. Moseley and D. E. Williams, "Oxygen Surface Species on Semiconducting Oxides", in Techniques and Mechanism in Gas Sensing , Edited by P. T. Mosely, J. O. W. Norris, D. E. William, Adam Hilger, Bristol, USA, p.46-60,1991.

70

[16] J. Maier, W. Goepel, "Investigations of bulk defect chemistry of polycrystalline Tin (IV) Oxide", *Journal of Solid State Chemistry*, 72, 293-302 (1988).

[17] W. Goepel, K.D. Schierbaum, "SnO_2 sensors: current status and future prospects", *Sensors and Actuators B*, 26-27, 1-12 (1995).

[18] R. Sorita, T. Kawano, "A Highly Selective CO Sensor, Screening of Electrode Materials", *Sensors and Actuators B*, 35-36, 274-277(1996).

[19] T. Kolodiazhnyi and A. Petric, "Effect of Reducing and Oxidizing Atmospheres on the PTCR Properties of $BaTiO_3$" in "Defects and Surface-Induced Effects in Advanced Perovskites " Edited by G. Borstel, A. Krumins, and D. Millers, NATO Science Series, 3. High Technology, Vol.77, Kluwer Academic Pub., p.467-472, 2000.

[20] S. R. Morrison, *The Chemical Physics of Surfaces*, Second Edition, Plenum Press, New York, p. 257, 1990.

[21] L. B. Kong, Y.S. Shen, "Gas-sensing Property and Mechanism of $Ca_xLa_{1-x}FeO_3$ Ceramics", *Sensors and Actuators B*, 30, 217-221(1996).

[22] P. D. Skafidas, P. S. Vlachos, J. N. Avantsiotis, "Modeling and Simulation of Abnormal Behavior of Thick-film Tin Oxide Gas Sensors in CO", *Sensors and Actuators B*, 21, 109-121 (1994).

[23] U. Hoefer, G. kuhner, W. Schweizer, G. Sulz, K. Steiner, "CO and CO_2 Thin-film SnO_2 Gas Sensor on Si Substrates", *Sensors and Actuators B*, 22, 115- (1994).

[24] Zhi-Gang Zhou, Zi-Long Tang, "Defects in Chemical Ceramic Sensors", in *"Encyclopedia of Sensors"*, Editors by C. A. Grimes, E. Dickey, and M. V. Pishko, will be published in American Scientific Publishers, 2005.

[25] E. Possenriede, H. Kröse, T. Varnhorst, R. Scharfschwerdt, "Shallow Acceptor and Electron Conduction States in $BaTiO_3$", *Ferroelectrics,* 151, 199-204 (1994).

[26] T. Varnhorst, O. F. Schirmer, H. Kröse, and R. Scharfschwerdt, "O^- holes associated with alkali acceptors in $BaTiO_3$", *Physical Review B,* 53 [1], 116-125 (1996).

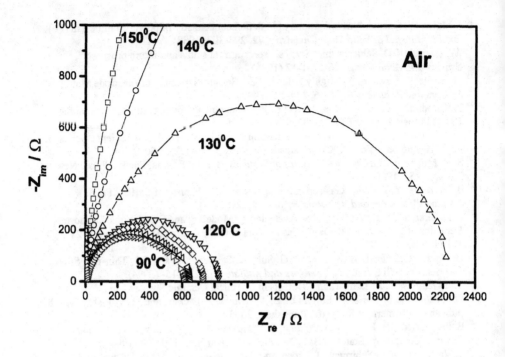

Fig. 1. Impedance spectra obtained for sample in air at different temperatures.
The two temperatures not noted are 100 and 110°C.

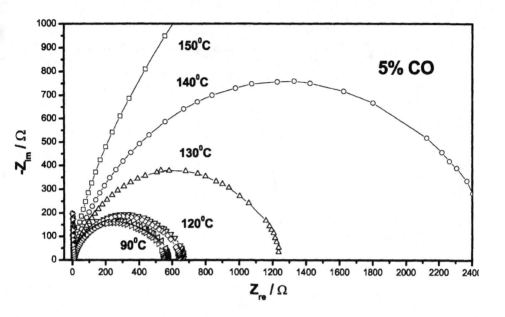

Fig. 2. Impedance spectra obtained for sample in 5 % CO gas at different temperatures. The two temperatures not noted are 100 and 110°C.

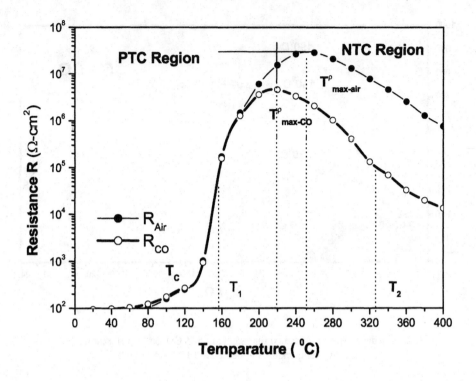

Fig. 3. PTCR characteristics in deferent surrounding atmosphere. Black and open circle, represented overall resistances in air and 1% CO gas, respectively. It is capable of detecting carbon monoxide in the PTC region at T_1 (below the maximum temperature) and in the NTC region at T_2 (above the maximum temperature). In which the $T^{\rho}_{max-air}$ and T^{ρ}_{max-CO} is the maximum resistance in air and CO gas, respectively.

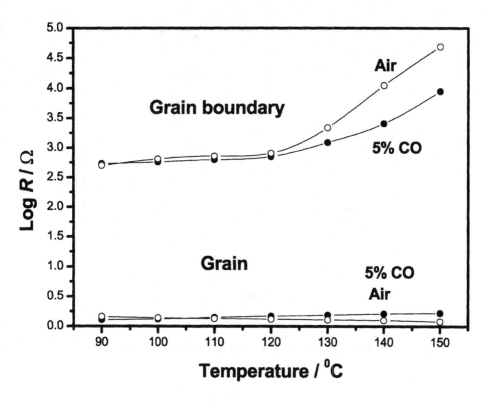

Fig. 4. Variation of R_{gb} and R_g are differenced with different temperature under differing atmosphere. The open circles represent data of sample working in air and the closed circles represent in 5 % CO gas.

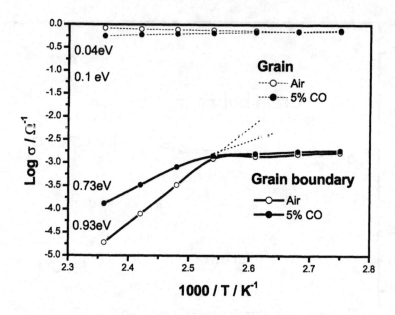

Fig. 5. Arrhenius plot of conductivity data for different atmosphere under differing temperatures. The open circles represent data of sample working in air and the closed circles represent in 5 % CO gas.

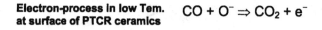

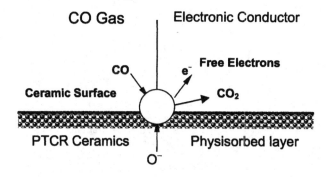

Fig. 6. Schematic illustrating the electron-process at a surface of PTCR ceramics as well as generation of free electrons in lower temperature.

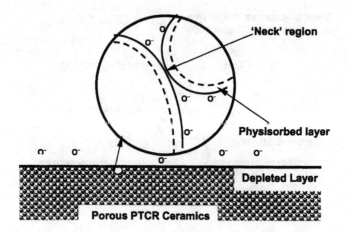

Fig. 7. Schematic diagram of a section through a porous PTCR-CO ceramics. The 'neck' region between the grains showing the effect of adsorbed oxygen.

FULL RANGE DYNAMIC STUDY OF EXHAUST GAS OXYGEN SENSORS

Da Yu Wang*, Eric Detwiler
Delphi
51786 Shelby Parkway
Shelby Township, MI, 48315, USA

ABSTRACT

We used transfer function approach to investigate exhaust oxygen sensor dynamics in engine exhaust environment. All the sensor dynamic mechanisms were identified. They are attributed to the effect of individual component of the sensor, such as the louver-shield, the protection coating-layer, and the sensing electrode; the gas transport system, such as the exhaust pipe gas blending effect; the exhaust generation system, such as the engine liquid fuel wall wetting, and the discrete individual cylinder event.

INTRODUCTION

Exhaust gas after-treatment systems rely on oxygen sensors for feedback loop control to minimize engine emissions. The system performance depends on the sensor dynamics. There were investigations on oxygen sensor dynamics, but the results were not conclusive.[1-4] Most of these studies failed to account for the fact that there are many dynamic mechanisms involved in an exhaust oxygen sensor. Fig. 1 list all the possible sensor dynamic contributors for both switching (Nernstian) oxygen sensor and wide range air to fuel ratio (amperometric) sensor.[6-8] Not that there is no negative proportional feedback loop for the electrode dynamic when a Nernstian oxygen sensor is referred in Fig. 1.[8] As shown in Fig. 1, unless all the dynamic effects are accounted for, it is difficult to compare and to draw conclusions from various sensor dynamic studies.

We used the transfer function approach to study these dynamics.[5] We had successes in the past using this approach to study the effects of the exhaust pipe, the sensor louver shield, the sensor coating layer and the sensor electrode on sensor response time.[6-8] Recently, good understandings of the effect of engine dynamics on exhaust gas have been obtained.[9-14] The wall-wetting effect of the fuel delivery has been established as the most important source for the transient air-fuel ratio excursions. In this study, we investigated these models and made a full range of sensor dynamic study by operating the sensors in engine exhaust environment.

THEORY

When fuel is injected into engine, some will stay airborne and the rest will wet the manifold walls, and evaporate again to participate next cycle of engine event. The transfer function of this process can be expressed in following equation.[8-13]

$$G_{ww}(z) = (1 - \beta) + [(1 - \alpha)z^{-1}\beta/(1 - \alpha z^{-1})] \qquad (1)$$

where α is the residual factor, the ratio of the residual fuel that remains on the manifold walls after a time interval of $\Delta t = t_k - t_{k-1}$ (which is 120/RPM [sec] for a four stroke combustion engine), β is the impact factor, the ratio of injected fuel wetting the wall, z is defined as,

79

$$z = e^{(i\omega\Delta t)}, \qquad (2)$$

denoting the complex shift operator, ω representing the angular frequency, and

$$\Delta t = 120 / RPM, \qquad (3)$$

representing the time of a full engine cycle in a four-stroke engine operating at a constant RPM.

Fuel evaporation is much more complex than a single process proposed in Eq. (1).[14] For a two-process approach, we will separate the fuel into two different portions C_1 and C_2, and each has individual residual factor, α_1 and α_2. The above equation then becomes,

$$G_{ww}(z) = (1 - \beta) + [(1 - \alpha_1)z^{-1}\beta/(1 - \alpha_1 z^{-1})]C_1 + [(1 - \alpha_2)z^{-1}\beta/(1 - \alpha_2 z^{-1})]C_2 \quad (4)$$

In the transfer function measurement, we input the frequency-sweep sine waves to the engine controller to generate the output exhaust gas. The engine individual cylinder event produces a stepwise sine-wave signal. This discrete-process gives a sample-and-hold dynamic effect on the exhaust gas.[5] The sample-and-holding transfer function is,

$$G_{ho}(s) = [1 - e^{(-Ts)}]/s \qquad (5)$$

where $s = i\omega$ and $T = 120/4/RPM$ for a four cylinder engine operated at a constant RPM.

The exhaust transport system shown in Figs. 1 is an exhaust pipe or an exhaust pipe with a converter canister attached. If the sensor is mounted ahead the canister, the pipe dynamic can be described by the plug flow (piston flow) reactor model and the transfer function is,

$$G(s) = e^{-s\theta}. \qquad (6)$$

where θ, is the transport lag in frequency spectrum.[6, 154-16] If the sensor is mounted behind the canister, the exhaust pipe can be described as the back-mixed flow (perfectly mixed flow) model and the transfer function is,

$$G(s) = [1/(s + \frac{1}{\tau})]e^{-s\theta}. \qquad (7)$$

where τ is the time constant (the volumetric rate of flow to fill the volume of the canister).[6, 15-16]

The exhaust gas sensors have louver shields and porous coating layers to protect the sensing electrodes from the exhaust soot poison effect. We have found that the louver shield is like a back-mixed flow reactor, and the transfer function is the same as Eq. (7).[6] Porous coating layers and the gas-diffusion-limiting apertures used by exhaust gas sensors can be modeled by a one-dimensional mass transport equation. The transfer function is,

$$G(s,u) = 1 - \sum(4/n\pi)[s \times \sin(n\pi u/2L)/[s + (n\pi/2L)^2 D], \qquad (8)$$

where $s = i\omega$, $u = L$, is the effective gas diffusion path length of the means of diffusion, and D is the gas diffusion constant.[7] The Nernstian type of sensor relies on the sensing electrode to sense the exhaust gas electrochemically (Nernst principle). The dynamic of an electrochemical electrode is determined by the two-dimensional surface diffusion of surface-adsorbed oxygen on the surface of the electrode. The transfer function is,

$$\overline{G}(s) = \sum 2(D/L^2)/[s + (n\pi/2)^2(D/L^2)] \qquad (9)$$

with $s = i\omega$, 2L is the average width of the electrode and D is the surface diffusion constant.[8]

The dynamics of engine controllers are fast and they are transparent to the sensor measurement. Signal filtering is frequently used in processing sensor output signals, as engine operations generate abundant noises. Typically, these filters are of the first order type, which can be described by Eq. (7). It is not uncommon to have more than one filter used in the signal output circuit boards and their cut-off frequencies can vary over a wide range. The transfer functions can be modeled by Eq. (7), or can be determined from experiments directly.

EXPERIMENTAL

We use a four cylinder 1.8L-2.4L engine with port fuel injection to deliver the exhaust gas. The fuel injectors are controlled by an electronic set point controller (PC ESC), or a dsp controller to create the air to fuel ratio perturbation. The frequency range of the swept sine extends from 0.01 Hz to 100 Hz. All tests are performed at either lean or rich engine condition. The air to fuel ratio set-point is specified by a Horiba MEXA-110 AFR analyzer. A HP3566A, Multi-channel Spectrum/Network Analyzer is used to provide the input of the sine sweep signals, to collect the output of the sensors. Before taking the data, the engines were warmed up for at least a half hour. Typically, the lean and rich conditions were set at least 1.5 to 2.0 air-fuel ratio from the stoichiometric point. Since sinusoidal fuel perturbations of $\pm(1\text{-}10\%)$ yielded similar results, most of the data were taken at $\pm5\%$.

RESULTS

The dynamic characteristics of the sensing devices are quite similar, regardless of the air to fuel ratio points tested. An example of a sensor gain responses is shown in Fig. 2, in which the sensor is an air to fuel ratio sensor operated at its amperometric mode. The corresponding phase response data are shown in Fig. 3. The testing frequencies are normalized to the RPM of the engine (1200 RPM) and the averaged air to fuel ratio is set at 16.5. The data can be fitted well with the transfer function models given in the theory section. The solid lines shown in Figs. 2 and 3 are the results of the modeling. Similar results were obtained for the same sensor when operated in Nernstian mode (we stopped the feedback loop control and used its emf cell to measure the sensor response). Again we can fit the data well with the transfer function models given in the theory section. Since the data were only slightly different from that shown in Figs. 2 and 3 (it contained electrode dynamic), only the data differences between the experiment and the modeling were shown in Fig. 4. The dark circles are the errors for the gain part of the data and the open squares are for the phase part of data. The parameters used in fitting the data obtained by operating the sensor in either pump mode (close feedback control) or Nernstian mode were listed in Table 1. Note that the same parameters were used in fitting the data, except the parameter of Eq. 9 (electrode transfer function) was used for the data of Nernstian mode. Also noted that the last four parameters to the right hand side of the Table 1 had been verified independently in previous studies.[6-8]

Table 1: Parameters used for the data shown in Figs. 2-6

Test Mode	β Eq. 4	C_1 Eq. 4	α_1 Eq. 4	C_2 Eq. 4	α_2 Eq. 4	θ Eq. 6	τ Eq. 7	$L/\sqrt{D}$ Eq. 8	$L/\sqrt{D}$ Eq. 9
Pump	51%	70%	3.6%	30%	72%	71 Deg/Hz	25 Hz	0.35 $sec^{1/2}$	None
Switch	51%	70%	3.6%	30%	72%	67 Deg/Hz	25 Hz	0.35 $sec^{1/2}$	0.40 $sec^{1/2}$

Figs 5 and 6 show the part of data that correspond only to the engine dynamic. We obtained the data by subtracting the contributions of Eqs. 5-9 from the measured transport function data, which were shown in Figs. 2 and 3. The data in Figs. 5 and 6 were fitted with the wall-wetting model. The solid curves represent the fitting results and the parameters used in this fitting already listed in Table 1.

DISCUSSION

The important result of this study is that we can account for all the dynamics involved in an exhaust gas oxygen sensor operated in engine exhaust environment. This statement is supported by the data shown in Figs. 2-6.

In Figs. 2-6, the errors become bigger as the normalized frequency goes beyond one. This is the cause of the discrete cylinder event. At the normalized frequency of one, a single periodic sine wave is described by four cylinder events. When the normalized frequency reaches four, there is only one cylinder event to describe a sine wave, which introduces the biggest error in Figs. 3-8 at this point.

In Fig. 2-6, there are larger errors near the normalized frequency of 0.5. The misfit of the data between the experiment and the model at this point is most likely the cause of the two-pole distribution of fuel evaporation process assumed in Eq. (4). Using a multi-polar distribution for the fuel evaporation process instead the bi-polar, one can improve the quality of the fittings further.

The dynamic effect of porous coating layers listed in Table 1 is significant. This is because the sensor is amperometric type, which requires the use of a thick diffusion-limiting aperture to create a limiting current when operated in pump mode. As indicated by Eq. 8, reducing the pore size and shortening the diffusion pathway can decrease the response time.

The electrode dynamic effect exists only in switching (Nernst) type oxygen sensors (see Eq. 9). The data indicates that it is as slow as that of the gas-diffusion-limiting aperture (see Table 1). Again, based on Eq. 9, a reduction of the diffusion length can shorten the response time.

The louver shield has fast frequency response as indicated in Table 1. A much slower dynamic effect can be easily realized by using a more restrictive louver shield design in the sensor package.

The flight delay time calculated from the θ values listed in Table 1 is around 0.19 second. This is reasonable considering the speed of the exhaust gas within the exhaust pipe and the distance the gas traveled between the engine exhaust outlet and the sensor.

Part of the parameters listed in Table 1 (the last four to the right hand of the table) had been verified independently by way of individual component study.[6-8] This was carried out by studying the difference of two transfer functions measured on two systems, with one included and the other one excluded the component.[6-8]

ACKNOWLEDGMENTS

The authors would like to thank Russ Bosh, Fred Kennard, Paul Kikuchi, Joachim Kupe, and Michel Sultan for their support and encouragement in publishing this work. Special thanks also go to Raymond C. Turin for many discussions and help he offered for this study. The authors particularly like to thank Linos Jacovides in providing the facilities and the environment at Delphi Research Laboratories to make this study possible.

REFERENCES

[1]S. P. S. Badwal, M. J. Bannister, F. T. Ciacchi, and G. A. Hooshmand, "Response rate techniques for zirconia-based nernstian oxygen sensors," *J. Appl. Electrochem.* 18, (1988) 608.

[2]M. M. Hedges and M. J. Bannister, Transient response of a low temperature zirconia oxygen sensor, *Key Engineering Materials* 53/54. 263-267 (1991).

[3]A. Sharma and P. D. Pacey, "*Rate and mechanism of response of a zirconia oxygen sensor,*" *J. Elecrochem. Soc.* 140, 2302-2309 (1993).

[4]C. T. Young, "Experimental analysis of ZrO2 oxygen sensor transient switching behavior, " *SAE* 810380, (1979).

[5]M. E. El-Hawary, Control System Engineering, Reston Publishing, Reston, Virginia, (1984).

[6]Da Yu Wang, Eric Detwiler, and Scott Nelson, Delphi internal report, "Dynamic study of louver shield and exhaust pipe for exhaust gas oxygen sensors," November 20, 2001.

[7]Da Yu Wang, Eric Detwiler, "Dynamic Study of Coating Layers of Exhaust Oxygen Sensors," *Sensor and Actuators B*, 106 (1) 229-233 (2005).

[8]Da Yu Wang and Eric Detwiler, "Electrode Dynamic Study of Exhaust Gas Oxygen Sensors," *Sensors and Actuators B*, 99 (2-3), 571-578, May 2004.

[9]C. F. Aquino, Transient A/F Control Characteristics of the 5 Liter Central Fuel Injection Engine, *SAE* 810494.

[10]Raymond C. Turin, Ph. D. Thesis No. 9999, "Untersuchung modellbasierter, adaptiver Verfahren zur Kompensation der Gemischbildungsdynamik eines Otto-Motors," Swiss Federal Institute of Technology (ETH), Zurich, 1992.

[11]Raymond C. Turin and Hans P. Geering, "On-line Identification of A/F Ratio Dynamics in a sequentially injected SI Engine," *SAE* 930857.

[12]Raymond C. Turin, Ernesto G. B. Casartelli, and Hans P. Geering, "A New Model for Fuel Supply Dynamics in an SI Engine," *SAE* 940208.

[13]Christopher H. Onder, "Modellbasierte Optimierung der Steuerung und Regelung eines Automobilmotors," Ph.D. Thesis No. 10323, Swiss Federal Institute of Technology (ETH), Zurich, 1993.

[14]W. O. Siegl, M. T. Guenther and T. Henney, "Identifying sources of evaporative emissions-using hydrocarbon profiles to identify emission sources," *SAE* 2000-01-1139.

[15]John R. Heywood, Internal Combustion Engine Fundamentals, McGraw-Hill Inc., New York, New York, 1988, p 210.

[16]Nickolas J. Themelis, Transport and Chemical Rate Phenomena, Gordon and Breach Publishers, Amsterdam, 1995.

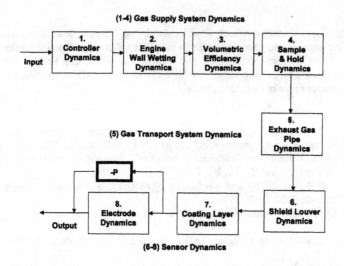

Fig. 1. The block diagram for the dynamics of a wide range air to fuel ratio sensor.

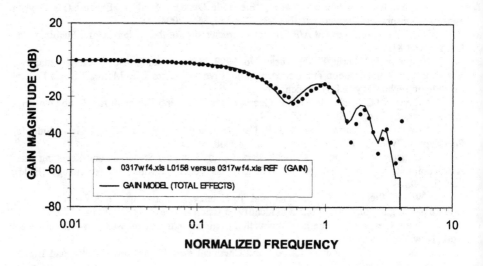

Fig. 2. Gain plot of the full frequency response of a wide range air to fuel ratio sensor operated in lean condition.

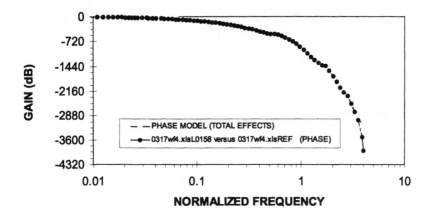

Fig. 3. The corresponding phase plot of Fig. 2.

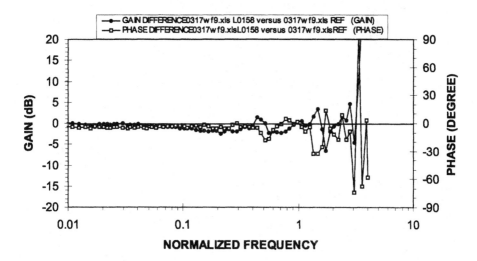

Fig. 4 Gain and phase plots of data difference between experiments and the fitting models of the same sensor as Fig. 2 but operated in switching mode in the same lean condition.

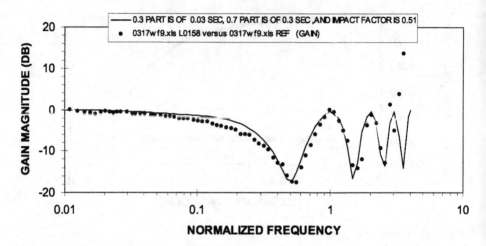

Fig. 5. The plot of wall-wetting transfer function (gain only) of the experimental data (dark circles) and the fitting model (dark curves). The sensor is a wide range air to fuel ratio sensor operated in switching mode in lean engine condition.

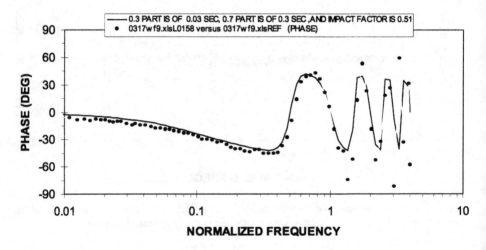

Fig. 6. The corresponding phase plot of Fig. 5.

Advanced Dielectric
Materials Phenomena

DIELECTRIC PROPERTIES OF NM-SIZED BARIUM TITANATE FINE PARTICLES AND THEIR SIZE DEPENDENCE

Satoshi Wada, Takuya Hoshina, Hiroaki Yasuno, Masanori Ohishi, Hirofumi Kakemoto, Takaaki Tsurumi and Masatomo Yashima
Tokyo Institute of Technology
2-12-1 Ookayama
Meguro, Tokyo, 152-8552

ABSTRACT

The powder dielectric measurement of barium titanate ($BaTiO_3$) fine particles with sizes from 17 to 1,000 nm revealed the maximum dielectric constant at a certain particle size. Moreover, the sizes with maximum dielectric constants were strongly dependent on the preparation methods. When the $BaTiO_3$ fine particles were prepared using the original 2-step thermal decomposition method, a dielectric constant maximum of 15,000 was observed at 70 nm particle size. On the other hand, when the $BaTiO_3$ fine particles were prepared using the modified 3-step thermal decomposition method, a dielectric constant maximum of 5,000 was observed at 140 nm. The former $BaTiO_3$ was prepared in vacuum of 10^{-2} torr while the latter $BaTiO_3$ was prepared in air. Structure refinement of $BaTiO_3$ particles using a Rietveld method revealed that all of $BaTiO_3$ particles were always composed of two parts; (a) surface cubic layer and (b) bulk tetragonal layer. Moreover, a thickness of surface cubic layer for $BaTiO_3$ nanoparticles prepared in vacuum of 10^{-2} torr was much thinner than that for $BaTiO_3$ nanoparticles prepared in air. Thus, to explain these differences, a new model on the basis of "surface relaxation" was proposed.

INTRODUCTION

Ferroelectric $BaTiO_3$ fine particles have been used as raw materials for electronic devices such as multilayered ceramic capacitors (MLCC). Recently, with the miniaturization of electronic devices, the down-sizing of MLCC has been developed and accelerated. As a result, it is expected that the thickness of dielectric layers in MLCC will become less than 0.5 µm. Consequently, the particle size of the required $BaTiO_3$ raw materials will decrease from a few hundred nm to a few tens of nm. However, in ferroelectric fine particles, it is known that ferroelectricity decreases with decreasing particle and grain sizes, and disappears below certain critical sizes; this is called the "size effect" in ferroelectrics.[1-8] Therefore, the size effect in ferroelectrics such as $BaTiO_3$ is one of the most important phenomena of interest with respect to the industry and science of ceramic capacitors. To date, many researchers investigated the size effect of $BaTiO_3$ using ceramics, but the size effect in $BaTiO_3$ ceramics is a very complicated phenomena because of coexistence of factors such as internal stress, grain boundaries, etc., except for grain size. Therefore, to investigate an intrinsic size effect, a study using $BaTiO_3$ single crystal particles should be required.

Recently, a new preparation method with crystal growth under vacuum (10^{-2} torr) to prepare defect-free, impurity-free $BaTiO_3$ nanoparticles was proposed by Wada et al.[9] Using this method, $BaTiO_3$ fine particles were successfully prepared from 17 nm to 500 nm. Moreover, they also developed a new measurement method for dielectric constants of $BaTiO_3$ particles.[10] A combination between this powder dielectric measurement method and the defect-free, impurity-free $BaTiO_3$ fine particles resulted in the unique particle size dependence with maximum dielectric constant of 15,000 at 68 nm.[11] On the other hand, Hoshina *et al.* modified the above preparation method to prepare defect-free, impurity-free $BaTiO_3$ nanoparticles with an amount over 60 g per 1 preparation[12] and in this method, a crystal growth was performed in air. Moreover, they also investigated a relationship between particle size and dielectric constant. As a result, the size dependence on powder dielectric constants exhibited a dielectric constant maximum of 5,000 at 140 nm.[12,13] This suggested that the size dependence was strongly dependent of the preparation methods. Therefore, if we can clarify an origin of the difference between the above two size particle dependences of dielectric constants, we can determine the underlying principles of the dielectric property - size effect for $BaTiO_3$ particles.

In this study, the objective is to clearly define an origin of the difference between the above two particle size dependences, and propose a new model to explain a size effect. For this objective, $BaTiO_3$ particles with various particle sizes from 17 to 1,000 nm were prepared using two different preparation methods from barium titanyl oxalate, *i.e.*, (1) the 2-step thermal decomposition method and (2) the modified 3-step thermal decomposition method, and their powder dielectric properties were measured. Moreover, the crystal structure of the $BaTiO_3$ particles was investigated using synchrotron XRD technique and the Rietveld method.

EXPERIMENTAL

Barium titanyl oxalate ($BaTiO(C_2O_4)_2 \cdot 4H_2O$) were prepared by Fuji Titanium Co., Ltd. Its Ba/Ti atomic ratio was 1.000 and the amount of the impurity was less than 0.02 mass%.[14] To prepare defect-free, impurity-free $BaTiO_3$ nanoparticles, two kinds of preparation methods, (1) the 2-step thermal decomposition method[9] and (2) the 3-step thermal decomposition method[12]. For the both method, the 1st step is the same way, and the thermal decomposition of $BaTiO(C_2O_4)_2 \cdot 4H_2O$ was performed at 500 °C for 3 hours in air, and resulted in the formation of the intermediate compounds ($Ba_2Ti_2O_5CO_3$) with almost amorphous structure. For the 2-step thermal decomposition method, the thermal decomposition of $Ba_2Ti_2O_5CO_3$ at the 2nd step was performed from 600 to 1,000 °C for 3 hours in vacuum of 10^{-2} torr. As a result, $BaTiO_3$ particles with various particle sizes ranging from 17 to 500 nm were prepared. On the other hand, for the 3-step thermal decomposition method, the thermal decomposition of $Ba_2Ti_2O_5CO_3$ at the 2nd step was performed at 650 °C for 3 hours in vacuum of 10^{-2} torr, and resulted in the formation of $BaTiO_3$ nanoparticles with 17 nm. At the 3rd step, the $BaTiO_3$ nanoparticles with 17 nm were annealed from 700 to 1,000 °C in air for the particle growth. As a result, $BaTiO_3$ particles with various particle sizes from 17 to 1,000 nm were prepared.

These particles were characterized using the following methods. The crystal structure was investigated using a powder X-ray diffractometer (XRD) (RINT2000, Rigaku, Cu-Kα, 50 kV, 30 mA). The average particle sizes and crystallite sizes were estimated using a transmission electron microscope (TEM) (CM300, Philips, 300 kV) and XRD. The impurity in the products was analyzed using a Fourier transform infrared spectrometer (FT-IR) (SYSTEM 2000 FT-IR, Perkin Elmer) and by differential thermal analysis with thermogravimetry (TG-DTA) (TG-DTA2000, Mac Science). The absolute density of the $BaTiO_3$ powders was measured using a pycnometer, and the relative density was calculated using a theoretical density estimated from lattice parameters by the XRD measurement. The Ba/Ti atomic ratios for the $BaTiO_3$ particles were determined by using the X-ray fluorescence analysis. The dielectric constants for these BaTiO₃ particles were measured by using the powder dielectric measurement method[10].

High intensity synchrotron XRD data were collected at beam line BL02B2 in the synchrotron radiation facility, SPring-8, JASRL. High energy X-ray with wavelength of 50.1049 pm was used as an incident X-ray. The BaTiO₃ powders were loaded into a glass capillary tube (0.2 mm in diameter) in vacuum and sealed. The diffraction patterns were recorded on the imaging plate with transmission geometry in the 2q ranges from 0.01° to 77.05°. Temperatures of samples were controlled by a N₂ gas flow system within ±1 °C, and were heated from 24 to 300 °C. This measurement was performed for all of the BaTiO₃ fine particles from 17 to 1,000 nm. Diffraction patterns were transformed from the Debye-Scherrer rings and analyzed by the Rietveld method using the TOPAS software (Bruker axs, version 2.1).

RESULTS AND DISCUSSION
Characterization and dielectric properties for BaTiO₃ particles

The obtained BaTiO₃ particles were characterized by using various methods. The details were described elsewhere.[11,12] As a result, the BaTiO₃ particles prepared by the both methods were defect-free, impurity-free particles, and their relative densities were over 99 % despite particle sizes. The particle sizes of the BaTiO₃ particles prepared by the 2-step thermal decomposition method ranged from 17 to 500 nm while those by the 3-step thermal decomposition method ranged from 17 to 1,000 nm. Especially, it should be noted that for the both methods, when the same annealing temperature was applied, a particle size of the BaTiO₃ particles prepared in vacuum was always a half of that of the BaTiO₃ particles prepared in air. This means that a growth mechanism for the BaTiO₃ particles may be different. For the both BaTiO₃ particles, the powder dielectric constants were measured at 20 ° C and 20 MHz.[10] Fig. 1 shows the results of powder dielectric measurement for the BaTiO₃ particles prepared in vacuum and air. From Fig. 1, the BaTiO₃ particles prepared in vacuum exhibited a maximum dielectric constant of 15,000 at 70 nm while those prepared in air possessed a maximum

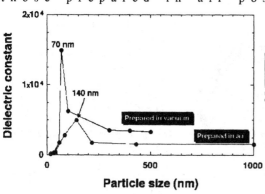

Particle size (nm)

Fig. 1. Particle size dependence of dielectric constants for the BaTiO₃ particles prepared in vacuum and air.

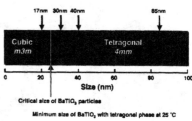

Fig. 2. Schematic diagram for particle size dependence of crystal symmetry for the BaTiO₃ particles prepared in vacuum and air.

91

dielectric constant of 5,000 at 140 nm. These results were completely consistent with the previous reports[11,12]. This difference between the both size dependences suggested that the size dependence of dielectric properties was strongly dependent of preparation methods and preparation conditions. Therefore, if we can find a responsible factor for this difference, we can determine a key point to understand the size effect for the $BaTiO_3$ particles. However, the previous characterization results did not give us any differences between the $BaTiO_3$ particles prepared in vacuum and air. Thus, we focus on the crystal structure refinement of these $BaTiO_3$ particles, and try to clarify some differences in their crystal structure.

Crystal symmetry assignment for $BaTiO_3$ particles

It is known that using conventional XRD equipment, it is difficult to assign the crystal symmetry of $BaTiO_3$ nanoparticles into either cubic or tetragonal symmetry owing to line broadening and low XRD intensity. To solve this problem, high intensity XRD patterns of the above $BaTiO_3$ particles were measured using synchrotron XRD technique. The maximum Miller index observed in this measurement was (077) plane with d_{077} of 40.616 pm, and using the higher Miller index, it is possible to refine the actual crystal structure. We must determine the crystal symmetry of the $BaTiO_3$ nanoparticles prior to the Rietveld analysis. The (002) and (200) planes of tetragonal $BaTiO_3$ particles becomes to one (200) plane above Curie temperature (Tc) while the (111) plane does not change through Tc. Thus, the temperature dependence in FWHM of the broad (111) and (200) planes was carefully measured and compared. Finally, we can determine the crystal symmetry at room temperature. On the basis of this concept, FWHM of the broad (111) and (200) planes for $BaTiO_3$ nanoparticles below 100 nm was measured at 24, 150 and 300 ˚C. As a result, for the both particles prepared in vacuum and air, the crystal symmetry at 17 nm was assigned to cubic Pm-$3m$ while the crystal symmetry over 30 nm was assigned to tetragonal $P4mm$ as shown in Fig. 2. Therefore, These results revealed that the critical size of $BaTiO_3$ particles, which is a size of ferroelectric phase transition from tetragonal to cubic at room temperature, exists between 17 and 30 nm.

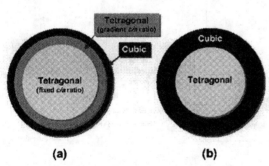

(a)   **(b)**

Fig. 3. Schematic models of the $BaTiO_3$ particle for the Rietveld fitting, (a) a three-phases model (one cubic phase and two tetragonal phases with fixed and gradient c/a ratio) and (b) a two-phases model (cubic and tetragonal phases).

Two-phases structures for $BaTiO_3$ particles

The Rietveld refinement was done for the data from 5˚ to 77˚ in 2q. In the fitting using a single tetragonal phase model, the final reliable factor (R_{wp}) and goodness of fitting (GOF) was 11.7 and 5.60, respectively. The GOF must be below 2.00 for the reliable fitting, but this GOF of 5.60 was too large. This large GOF was originated from the significant difference around (00l) and (h00) planes between measured and calculated values. Especially, in the

92

measured XRD profile, the unknown bridge structure between (00l) and (h00) planes was always observed, and this part caused the large difference between measured and calculated values. With decreasing particle size, the bridge structure between (00l) and (h00) planes became larger. To obtain good GOF below 2.00, the Rietveld fitting was performed using various models such as a two-phases model (cubic and tetragonal phases)[15] and a multi-phases model (one cubic phase and several tetragonal phases with different c/a ratios)[16]. As a result, it was found that when the three-phases model (one cubic phase and two tetragonal phases with fixed and gradient c/a ratios) was used as shown in Fig. 3-(a), the GOF closed to 1.00. This suggested that in the BaTiO$_3$ particles, there is intrinsically the region with gradient c/a ratios from 1.011 to 1.00. At present, we cannot assign this gradient region to (1) surface, (2) domain wall and (3) both surface and domain wall. Moreover, it is difficult to induce the gradient region into a model for the Rietveld analysis. Thus, we used a simple two-phases model (cubic and tetragonal phases) as shown in Fig. 3-(b) for the Rietveld fitting and analyzed the crystal structure for the BaTiO$_3$ particles from 17 to 1,000 nm. At present, we believe that on the basis of the surface relaxation model reported by Ishikawa[17], all of the BaTiO$_3$ particles should be composed of two parts, (a) surface cubic layer and (b) bulk tetragonal layer. The Rietveld analysis with the two-phases model gave us very important information such as (1) tetragonal/cubic volume ratio, (2) lattice parameters of cubic and tetragonal phases, and (3) ion's position in each phase. Fig. 4 shows the particle size dependence of the tetragonal/cubic volume ratio for the both particles prepared in vacuum and air. From Fig. 4, the tetragonal/cubic volume ratio in the BaTiO$_3$ particles prepared in vacuum is always twice larger value than that prepared in air. Moreover, using the tetragonal/cubic volume ratio and particle size, we can calculate the thickness of the surface cubic layer. Fig. 5 shows the thickness of the surface cubic layer for the both particles prepared in vacuum and air. From Fig. 5, this thickness for the BaTiO$_3$ particles prepared in vacuum is always a half of that prepared in air. Here, it should be noted that the cubic phase in this surface layer is very special, and is quite different from normal cubic phase of BaTiO$_3$. This is because surface layer can be distorted for a relaxation of surface tension, and this surface structure cannot change to another structure with changing t e m p e r a t u r e .   M o r e o v e r ,   p o w d e r   d i e l e c t r i c

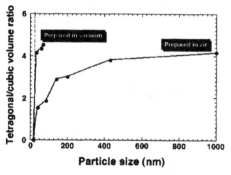

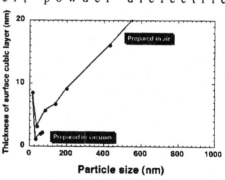

Fig. 4. Particle size dependence of tetragonal/cubic volume ratio for the BaTiO$_3$ particles prepared in vacuum and air.

Fig. 5. Particle size dependence of surface cubic layer thickness for the BaTiO$_3$ particles prepared in vacuum and air.

measurement of the BaTiO$_3$ particles with 17 nm size resulted in a low dielectric constant of around 230. Therefore, now we believe that surface cubic layer is a very special cubic phase with a low dielectric constant of 230. This result (Fig. 5) and dielectric measurement results (Fig. 1) also revealed that for the BaTiO$_3$ particles with the same particle size, as the thickness of the surface cubic layer becomes thin, the dielectric constant becomes larger. Thus, the tetragonal/cubic volume ratio is one of the most responsible factors for both the higher dielectric constants and provides an explanation of the size effect.

Crystal structure for tetragonal bulk layer

The crystal structure of the bulk tetragonal layer was also refined by the Rietveld fitting. Fig. 6 shows the particle size dependence of the lattice parameters for the BaTiO$_3$ particles prepared in vacuum and air. From Fig. 6, for the BaTiO$_3$ particles prepared in air, with decreasing particle sizes, the c-axis slightly decreased while above 200 nm the a-axis was almost constant and below 200 nm, the a-axis significantly increased. On the other hand, for the BaTiO$_3$ particles prepared in vacuum, with decreasing particle sizes, the a-axis significantly increased while above 30 nm, the c-axis increased and below 30 nm, the c-axis slightly decreased. Now, for these BaTiO$_3$ particles, it is very difficult to explain the differences of these particle size dependences. Fig. 7 shows the particle size dependence of the tetragonality (c/a ratio) for the BaTiO$_3$ particles prepared in vacuum and air. For both particles, the c/a ratio decreased with decreasing particle sizes. The most important thing is that the maximum dielectric constant was obtained for a c/a ratio of 1.0046 in vacuum and 1.0077 in air, and as compared to these behaviors, the BaTiO$_3$ particles with lower c/a ratio exhibited much higher dielectric constant. This observation is very important in truly understanding the size effect.

Phonon analysis for BaTiO$_3$ particles

The dielectric properties of dielectric materials such as BaTiO$_3$ are always determined by phonon behavior, especially soft mode behavior. However, it is well-known that there is a significant difficulty to observe soft mode of BaTiO$_3$ directly. Thus, in this study, Raman

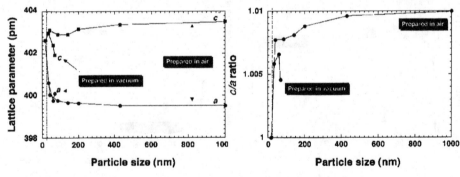

Fig. 6. Particle size dependence of lattice parameters for the BaTiO$_3$ particles prepared in vacuum and air.

Fig. 7. Particle size dependence of c/a ratio for the BaTiO$_3$ particles prepared in vacuum and air.

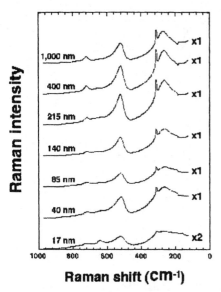

Raman shift (cm⁻¹)

Fig. 8. Particle size dependence of Raman scattering spectra for the BaTiO₃ particles prepared in air.

scattering spectra from 50 to 1,200 cm⁻¹ were measured for the BaTiO₃ particles prepared in vacuum and air. Fig.8 shows the particle size dependence of the Raman scattering spectra for the BaTiO₃ particles prepared in air. From this measurement, the optical phonon peaks except for the soft mode were observed, and each optical phonon mode was separated by using a peak fitting software (Galactic Ind., GRAMS/32 AI Ver.6, Gaussian type function) on the basis of the four kinds of optical phonon modes for the tetragonal *4mm* BaTiO₃ crystal as shown in Fig. 9. Finally, for the optical phonon modes, a resonance frequency was determined by peak position while the damping factor was determined by a FWHM of Raman peak. Fig. 10 shows the size dependence of the resonance frequency for the BaTiO₃ particles prepared in air. Despite particle sizes, the resonance frequencies of each optical mode were almost constant except for the soft mode. Fig. 11 shows the size dependence of the damping factor for the BaTiO₃ particles prepared in air. With decreasing particle sizes, the damping factor slightly increased, and at a certain size between 17 and 40 nm, all of damping factors significantly increased. The similar behavior was observed for the BaTiO₃ particles prepared in vacuum. This phenomenon was strongly related to the size-induced phase transition between cubic and tetragonal observed by a synchrotron XRD measurement.

For the normal BaTiO₃ single crystal, the temperature-induced phase transition behavior between cubic and tetragonal was explained by the softening of the soft mode (Slater mode for BaTiO₃)[18]. On the other hand, it is known that the resonance frequency of other optical modes

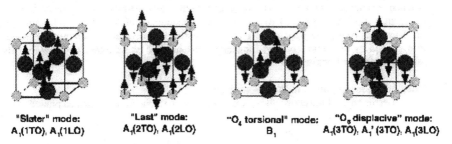

"Slater" mode: $A_1(1TO)$, $A_1(1LO)$ "Last" mode: $A_1(2TO)$, $A_1(2LO)$ "O_4 torsional" mode: B_1 "O_6 displacive" mode: $A_1(3TO)$, $A_1'(3TO)$, $A_1(3LO)$

Fig. 9. Schematic four kinds of optical phonon modes for the tetragonal *4mm* BaTiO₃ crystal.

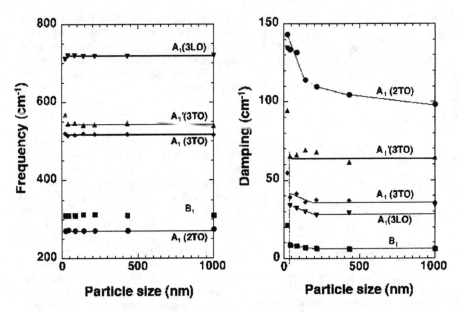

Fig. 10. Particle size dependence of optical phonon resonance frequencies for the BaTiO₃ particles prepared in air.

Fig. 11. Particle size dependence of optical phonon damping factors for the BaTiO₃ particles prepared in air.

for BaTiO₃ single crystal did not change significantly at Tc of 132 °C.[19] In this study, the temperature dependence of resonance frequency and damping factor was measured for some optical modes except for the soft mode of BaTiO₃ particles prepared in vacuum and air. As a result, despite temperature, the resonance frequency was almost constant while the damping factor slightly increased with increasing temperature, and significantly increased at Tc. In this study, we did not observe the soft mode, but near Tc, the resonance frequency of soft mode might reduce. Therefore, near Tc, the sudden increase of the damping factor means the the soft mode has further softened. On the basis of this discussion, we believe that the drastic increase of the damping factor at a particle size between 17 and 40 nm suggested the softening of the soft mode as shown in Fig. 12. This revealed that the sized-induced phase transition between cubic and tetragonal phases can be explained by the same model as the temperature-induced phase transition behavior between cubic and tetragonal phases.

A new model for size effect of BaTiO₃ particles

On the basis of the above discussion, we considered that the size-induced phased transition can be caused by the softening of the soft mode. This leads to the novel idea that the dielectric constant must indicate a maximum value (30,000~40,000) at a critical size, similar to the observation of maximum dielectric constant at Tc. Fig. 13 shows the schematic particle size dependence of dielectric constant on the basis of the above model. As a comparison, the results

Ferroelectric phase transition caused by size
(size effect on ferroelectrics)

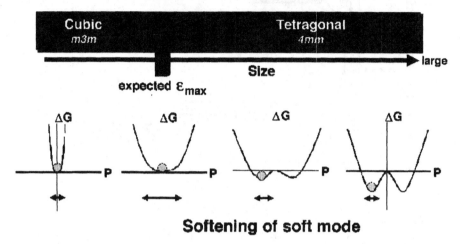

Softening of soft mode

Fig. 12. Schematic model for softening of soft mode caused by the down-sizing of the BaTiO₃ particles.

obtained in this study were also shown in Fig. 13. The ideal dielectric behavior on the basis of the above model shows a dielectric maximum of around 35,000 at a critical size between 17 and 30 nm. However, the results in this study exhibited the dielectric maximum at much higher particle sizes. We must explain this difference between three dielectric behaviors in Fig. 13. As previously mentioned, the key point for this difference should be a thickness of the surface cubic layer. It should be noted that for the ideal dielectric behavior on the basis of the model, there is no surface cubic layer. Moreover, as the thickness of the surface cubic layer increased, dielectric maximum values became lower and the particle size for the dielectric maximum increased. If the size dependence of dielectric properties for the BaTiO₃ particles with much thick surface cubic layer are measured, the dielectric maximum may not be observed, and dielectric constant may decrease with decreasing particle sizes. To date, there were many kinds of reports for the size dependence of the BaTiO₃ particles and ceramics. Some researchers reported the observation of the dielectric maximum[2,5] while other researches reported no observation of the dielectric maximum[1,4,7,8]. Now, we believe that key point is a thickness of the surface cubic layer, and this thickness strongly depends on the preparation methods and preparation conditions. In this study, we used the 2-step thermal decomposition method including the thermal decomposition and particle growth in vacuum, and prepared the BaTiO₃ particles with very thin surface cubic layer below 2 nm. This is a reason why we can observe the particle size dependence with a dielectric maximum. The particle growth rate in vacuum was almost half of that in air in this study, which indicated that this slow particle growth in clean atmosphere can be an origin of very thin surface cubic layer.

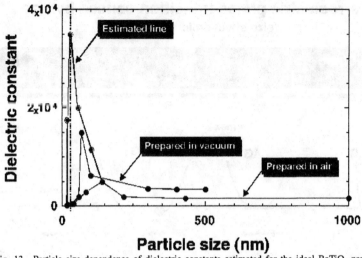

Fig. 13. Particle size dependence of dielectric constants estimated for the ideal BaTiO₃ particles without surface cubic layer, and measured using the BaTiO₃ particles prepared in vacuum and air.

Direction to obtain BaTiO₃ nanoparticles with ultrahigh dielectric constants

To date, a common sense for the size effect of the BaTiO₃ particles and ceramics is that high c/a ratio near 1.011 must be required for high dielectric constant. However, this study proposed an opposite idea that the lower c/a ratio closed to 1.0 must be required for high dielectric constant. Moreover, as another important point, a new factor of the thickness of the surface cubic layer was proposed to obtain high dielectric constant. If possible, an ideal thickness for the surface cubic layer is considered as one unit cell thickness. Fong *et al.* reported that the ferroelectricity of lead titanate film on strontium titanate substrate can be introduced over just three unit cells thickness, and one unit cell is enough to relax the surface tension if very clean surface is prepared[20]. Therefore, we believe that for the BaTiO₃ particles, if an ideal clean surface is prepared, one unit cell may be enough to relax the surface tension, i.e., the thickness of the ideal surface cubic layer is one unit cell thickness of around 0.4 nm. On the basis of the above discussion, we try to propose a new direction to obtain BaTiO₃ nanoparticles with ultrahigh dielectric constant over 30,000.

First, defect-free, impurity-free BaTiO₃ particles with relative densities over 99 % must be prepared. Second, for these BaTiO₃ particles, the thickness of the ideal surface cubic layer should be close to one unit cell thickness (around 0.4 nm). Third, satisfying the above two conditions by controlling only particle sizes, particles of the critical size, i.e., the size-induced phase transition size between cubic and tetragonal phases at room temperature, are obtained. This means that the c/a ratio for the bulk tetragonal layer in the BaTiO₃ particles approaches 1.0 keeping the tetragonal phase. If the above three conditions are satisfied, the BaTiO₃ nanoparticles with ultrahigh dielectric constant over 30,000 can be prepared. At present, on the basis of this concept, we are attempting to prepare BaTiO₃ nanoparticles with ultrahigh dielectric constants.

CONCLUSIONS

Two different preparation methods using barium titanyl oxalate were applied to obtain $BaTiO_3$ particles. The $BaTiO_3$ particles prepared in vacuum and air were characterized by various methods, and it was confirmed that these particles were defect-free, impurity-free $BaTiO_3$ particles with relative densities over 99 %. Moreover, particle sizes ranged from 17 nm to 1,000 nm. The powder dielectric measurement for these $BaTiO_3$ particles revealed that a dielectric constant maximum of 15,000 was observed at 70 nm for the $BaTiO_3$ fine particles prepared in vacuum of 10^{-2} torr while the dielectric constant maximum of 5,000 was observed at 140 nm for the $BaTiO_3$ fine particles prepared in air. To clarify these differences for dielectric properties, the structure refinement of $BaTiO_3$ particles using a Rietveld method was performed for all of the $BaTiO_3$ particles. As a result, all of the $BaTiO_3$ particles were always composed of two parts; (a) surface cubic layer and (b) bulk tetragonal layer. Moreover, a thickness of surface cubic layer for $BaTiO_3$ particles prepared in vacuum was much thinner than that for the $BaTiO_3$ particles prepared in air. Thus, to explain these differences, a new model on the basis of "surface relaxation" was proposed. On the other hand, Raman scattering measurements for the $BaTiO_3$ particles resulted in a similar vibration behavior of optical phonons, except for soft mode, between size-induced and temperature-induced phase transitions. This result suggested that the size-induced phase transition can be originated from a softening of the soft mode frequency. On the basis of these findings, a new model to explain the intrinsic size effect was proposed. To confirm this model, the size dependence of the soft mode frequency must be measured, which is a significant technical challenge.

ACKNOWLEDGMENTS

We would like to thank Mr. M. Nishido of Fuji Titanium Co., Ltd. for preparing high purity barium titanyl oxalates and Mr. K. Abe of Sakai Chemical Industry Co., Ltd. for providing high purity BT-05 powders. We also would like to thank Dr. K. Kato and Dr. M. Takata of SPring-8, JASRI for helpful discussion of the high intensity XRD measurement using the synchrotron radiation technique. The experiment at SPring-8 was carried out under Program No. **2004A0566-ND1d-np**. This study was partially supported by (1) a Grant-in-Aid for Scientific Research (**15360341**) from the Ministry of Education, Science, Sports and Culture, Japan and (2) the Ookura Kazuchika Memorial foundation.

REFERENCES

[1]K. Kinoshita and A. Yamaji, "Grain-size Effects on Dielectric Properties in Barium Titanate Ceramics," *J. Appl. Phys.*, **45**, 371-373 (1976).

[2]G. Arlt, D. Hennings and G. De With, "Dielectric Properties of Fine-grained Barium Titanate Ceramics," *J. Appl. Phys.*, **58**, 1619-1625 (1985).

[3]K. Ishikawa, K. Yoshikawa and N. Okada, "Size Effect on the Ferroelectric Phase Transition in PbTiO3 Ultrafine Particles," *Phys. Rev. B*, **37**, 5852-5855 (1988).

[4]K. Uchino, E. Sadanaga and T. Hirose, "Dependence of the crystal structure on particle size in barium titanate," *J. Am. Ceram. Soc.*, **72**, 1555-1558 (1989).

[5]M. H. Frey and D. A. Payne, "Grain-size Effect on Structure and Phase Transformations for Barium Titanate," *Phys. Rev. B*, **54**, 3158-3168 (1996).

[6]S. Wada, T. Suzuki and T. Noma, "Role of lattice defects in the size effect of barium titanate fine particles: A new model," *J. Ceram. Soc. Jpn.*, **104**, 383-392 (1996).

[7]D. McCauley, R. E. Newnham and C. A. Randall, "Intrinsic size effects in a BaTiO$_3$ glass ceramic," *J. Am. Ceram. Soc.*, **81**, 979-987 (1998).

[8]Z. Zhao, V. Buscaglia, M. Viviani, M. T. Buscaglia, L. Mitoseriu, A. Testino, M. Nygren, M. Johnsson and P. Nanni, "Grain-size Effect on the Ferroelectric Behavior of Dense Nanocrystalline BaTiO$_3$ Ceramics," *Phys. Rev. B*, **70**, 024107 (2004).

[9]S. Wada, M. Narahara, T, Hoshina, H. Kakemoto and T. Tsurumi, "Preparation of nm-sized BaTiO$_3$ Fine Particles Using a New 2-step Thermal Decomposition of Barium Titanyl Oxalates," *J. Mater. Sci.*, **38**, 2655-2660 (2003).

[10]S. Wada, T. Hoshina, H. Yasuno, S.-M. Nam, H. Kakemoto and T. Tsurumi, "Preparation of nm-sized BaTiO$_3$ Crystallites by The 2-step Thermal Decomposition of Barium Titanyl Oxalate and Their Dielectric Properties," *Key Eng. Mater.*, **248**, 19-22 (2003).

[11]S. Wada, H. Yasuno, T. Hoshina, S.-M. Nam, H. Kakemoto and T. Tsurumi, "Preparation of nm-sized barium titanate fine particles and their powder dielectric properties, " *Jpn. J. Appl. Phys.*, **42**, 6188-6195 (2003).

[12]T. Hoshina, H. Yasuno, S.-M. Nam, H. Kakemoto, T. Tsurumi and S. Wada, "Size Effect on Dielectric Properties of Barium Titanate Fine Particles," *Trans. Mater. Res. Soc. Jpn.*, **29**, 1207-1210 (2004).

[13]S. Wada, T. Hoshina, H. Yasuno, S.-M. Nam, H. Kakemoto and T. Tsurumi, "Origin of Ultrahigh Dielectric Properties of nm-sized Barium Titanate Crystallites," *Ceram. Trans.*, (2004) in press.

[14]T. Kajita and M. Nishido, "Preparation of Submicron Barium Titanate by Oxalate Process," *Extended Abstracts of the 9th US-Japan Seminar on Dielectric and Piezoelectric Ceramics*, Okinawa 425-427 (1999).

[15]S. Aoyagi, Y. Kuroiwa, A. Sawada, I. Yamashita and T. Atake, "Composite Structure of BaTiO$_3$ Nanoparticle Investigated by SR X-ray Diffraction," *J. Phys. Soc. Jpn.*, **71**, 1218-1221 (2002).

[16]T. Hoshina, H. Kakemoto, T. Tsurumi, S. Wada, M. Yashima, K. Kato and M. Takata, "Analysis of Composite Structures on Barium Titanate Fine Particles using Synchrotron Radiation," *Key Eng. Mater.*, (2005) in press.

[17]K. Ishikawa and T. Uemori, "Surface Relaxation in Ferroelectric Perovskite," *Phys. Rev. B*, **60**, 11841-11845 (1999).

[18]J. C. Slater, "The Lorentz Correction in Barium Titanate," *Phys. Rev.*, **78**, 748-761 (1950).

[19]S. Wada, T. Suzuki, M. Osada, M. Kakihana and T. Noma, "Change of Macroscopic and Microscopic Symmetry of Barium Titanate Single Crystal around Curie temperature," *Jpn. J. Appl. Phys.*, **37**, 5385-5393 (1998).

[20]D. D. Fong, G. B. Stephenson, S. K. Streiffer, J. A. Eastman, O. Auciello, P. H. Fuoss and C. Thompson, "Ferroelectricity in Ultrathin Perovskite Films," *Science*, **304**, 1650-1653 (2004).

THE EFFECT OF STARTING POWDERS ON THE GIANT DIELECTRIC PROPERTIES OF THE PEROVSKITE CaCu$_3$Ti$_4$O$_{12}$

Barry Bender and Ming-Jen Pan
U.S. Naval Research Laboratory
Code 6351
Washington, DC 20375

ABSTRACT

Three different starting powders were used to test the effect of sintering time at 1100°C on the dielectric properties of CaCu$_3$Ti$_4$O$_{12}$. Dielectric properties were correlated with the grain size microstructure using a standard IBLC model. The presence of excess CuO led to discontinuous grain growth with large flat porous grains. CaCu$_3$Ti$_4$O$_{12}$ when synthesized using anatase as a titania substitute for rutile showed signs of enhanced sintering which reduced the porosity of the large grains resulting in a significant drop in dielectric loss. Room temperature permittivity varied from 10,538 to 368,970 depending on sintering time and starting powder.

INTRODUCTION

The Navy is engaged in a future naval capability program to develop the all electric ship. Substantial improvements have been made in power control and conditioning. However, in the state of the art power converter, capacitors are the limiting factor in improved volumetric efficiency. These capacitors not only need to be smaller (greater permittivity), but have to be stable over a wide range of temperatures and operating voltages. Commercial dielectric oxides such as BaTiO$_3$ typically satisfy two of the conditions but not all three. Recently, a new dielectric oxide, CaCu$_3$Ti$_4$O$_{12}$, has been uncovered with the potential to have high permittivity (single crystal dielectric constant is 80,000) that is stable over a wide range of temperatures.[1,2] However, it appears that the dielectric properties and microstructure of CaCu$_3$Ti$_4$O$_{12}$ (CCTO) are very sensitive to processing. Dielectric constants from 478 to 300,000 have been measured for polycrystalline CaCu$_3$Ti$_4$O$_{12}$.[3,4] Variations in microstructure has been reported too. For example, Li et al.[5] measured a grain size of about 10 microns for a CCTO powder sintered for 12 h at 1100°C, while Bender et al.[6] observed abnormal grain growth in CCTO after it had been sintered for only 3 h at 1100°C with grain as large as 170 microns.

The reasons for the observed wide variation in dielectric properties for CCTO are unknown. This is because the nature of its giant permittivity is still open to scientific debate. Combined modeling and experimental research on CCTO indicate that the mechanism for the giant permittivity is not intrinsic.[4,7] The extrinsic mechanism most commonly believed to be the source of CCTO's high permittivity is the formation of internal capacitive barrier layers.[2,4,8] It is believed that during processing of CCTO that insulating surfaces form on semiconducting grains creating an electronically heterogeneous material very similar in nature to internal barrier layer capacitors (IBLC). The dielectric properties and microstructure of these IBLCs are very sensitive to different processing parameters.[9] This trend was observed in our earlier reported research where the room temperature dielectric constant of CCTO varied from 700 to 930,000 and the grain size varied from 2 to 170 microns depending on how it was processed.[6] Pan et al.[10] found that the dielectric properties are sensitive to the size and volume fraction of the large grains present in the dielectric ceramic. It was also observed that permittivity increased with increased sintering time and sintering temperature but often at the cost of an increase in dielectric loss.[6]

This paper reports on the use of three different CCTO powders to explore the effect of processing on the dielectric properties and microstructure of CCTO. The first powder synthesized from standard precursors was used to explore in detail the relationships between increased sintering time, microstructure, and dielectric properties. The second powder used anatase instead of rutile as the starting titania precursor in order to explore the possibility of observing the effects of faster sintering kinetics of CCTO on its microstructure and dielectric properties. The last powder was acid-treated to test the effect of CuO-free starting powder on abnormal grain growth in CCTO and its effect on permittivity.

EXPERIMENTAL PROCEDURE

$CaCu_3Ti_4O_{12}$ was prepared using conventional ceramic solid state reaction processing techniques. Stoichiometric amounts of $CaCO_3$ (99.98%), CuO (99.5%) and TiO_2 (99.5%- rutile or anatase) were mixed by attrition milling. The three powders were mixed by blending the precursor powders into a purified water solution containing a dispersant (Tamol 901) and a surfactant (Triton CF-10). The resultant slurries were then attrition-milled for one h and dried at 90°C. The standard processed powder, CSTD, was calcined for 8 h at 1000°C then 930°C for 4 h and 970°C for 4 h. The anatase-based powder, CANT, was calcined for 900°C for 4 h and then 945°C for 4 h. After the final calcination the CSTD and CANT powders were attrition-milled again for 1 h to produce finer powders. For the CuO-free powder, CCUF, the CSTD powder before its final attrition milling was soaked in 3 M HCl in order to digest any free CuO. A 2% PVA binder solution was mixed with the powder and the powder was sieved to eliminate any large agglomerates. The dried powder was uniaxially pressed into discs typically 13 mm in diameter and 1.5 mm in thickness. The discs were then placed on platinum foil and sintered in air at 1100°C for various dwell times (2 to 16 h).

Material characterization was done on the disc and powders after each processing step. X-ray diffraction (XRD) was used to monitor phase evolution for the various mixed powders and resultant discs. Microstructural characterization was done on the fracture surfaces using scanning electron microscopy (SEM). To measure the dielectric properties, sintered pellets were ground and polished to achieve flat and parallel surfaces onto which palladium-gold electrodes were sputtered. The capacitance and dielectric loss of each sample were measured as a function of temperature (-50 to 100 °C) and frequency (100 Hz to 100 KHz) using an integrated, computer-controlled system in combination with a Hewlett-Packard 4284A LCR meter.

RESULTS

A bimodal distribution of grain sizes developed when the standard CCTO powder, CSTD, was sintered for various times at 1100°C. Figure 1 shows the development of the CCTO microstructure from a mixture of small and large grains to all large grains with an increase in sintering time from 2 h to 16 h. As Fig. 2 shows the volume of small grains decreased steadily with increasing sintering time until at 12 h the CCTO microstructure consisted of all large grains in the range of 50 to 100 microns. The average grain size of the small grains was 2.7 microns and did not change with increasing sintering time. The relative density of CCTO did not change much either with sintering time as all the samples varied from 94.3 to 94.7% except for the sample sintered at 2h which had a density of 93.3%. All the large grains typically had pores 3 to 5 microns in size (see Fig. 1(g)), which is typical of grains undergoing abnormal grain growth. With increasing sintering time there appeared to be an increase in the number of grains that showed the presence of an unknown grain boundary phase and sometimes the presence of grain

102

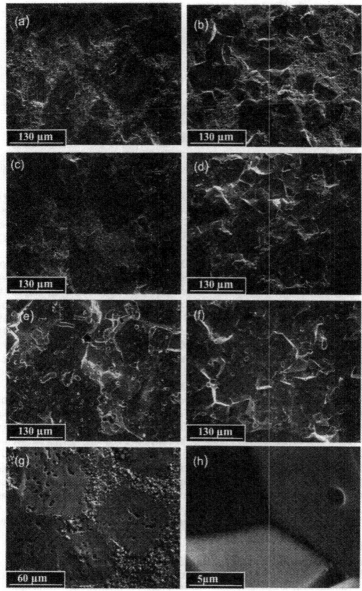

Fig. 1. SEM fractographs of CSTD powders sintered at 1100°C for (a) 2 h, (b) 3 h, (c) 4 h, (d) 8 h, (e) 12 h, and (f) 16 h. Fractograph (g) shows the presence of abnormal grains in a fine-grain matrix (2 h) and (h) shows the presence of a grain boundary phase and precipitates (8 h).

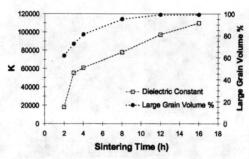

Fig. 2. The effect of sintering time on the permittivity (measured at 1 kHz at room temperature) and large grain volume percent of CCTO ceramics sintered from CSTD powders at 1100°C.

boundary precipitates (Fig. 1(h)). Figure 2 also shows the effect of sintering time of CSTD at 1100°C on the room temperature permittivity (1 kHz) of CCTO. The data shows a substantial increase in dielectric constant in sintering from 2 h to 3 h and after that a steady increase in permittivity with an increase in sintering time. After 16 h of sintering at 1100°C CaCu₃Ti₄O₁₂ has a room temperature dielectric constant of 109,790 and a dielectric loss of 0.053.

A bimodal distribution of grain size developed when the anatase-based CCTO powder, CANT, was sintered at 1100°C for 3 h (Fig. 3(a)). The volume of small grains was 27% and the average grain size was 1.7 microns and the sample had a relative density of 96.2%. The larger grains were not porous and were in the range of 5 to 60 microns in size (Fig. 3(d)). With longer sintering times (16 h) the bimodal distribution of grain size disappeared (see Fig. 3(b)) and the CCTO dielectric ceramic now showed the presence of only large grains in the range of 10 to 60 microns. These grains were also not porous but the relative density dropped to 93.0%. Figure 4 shows the temperature dependence of the permittivity and dielectric loss measured at 1 kHz with increasing sintering time at 1100°C (from 3 to 16 h). The dielectric constant and loss decreased when sintering time increased from 3 h (20749, 0.028) to 16 h (15241, 0.010).

No substantial abnormal grain growth developed when the acid-treated CSTD powder, CCUF, was sintered at 1100°C for either 3 or 16 h (Fig. 5(a&b)). The grain size and relative density increased with increasing sintering time from 2.7 microns to 3.7 microns and 89.0% to 94.0% respectively. Both samples did show the presence of a few large grains that were typically 10 to 20 microns in size but were not as porous as the large grains in the CSTD ceramics. No obvious grain boundary phases were apparent but the presence of titania grains was observed. Figure 4 shows the dielectric properties of these two samples as a function of temperature. The room temperature dielectric constant of the CCUF powder increased by an order of magnitude with increasing sintering time jumping from 10538 after 3 h at 1100°C to 361620 after 16 h. The dielectric loss also jumped going from 0.052 after 3h to 1.163 after 16 h.

DISCUSSION
The Effect of Sintering Time on the Permittivity of CCTO
Increasing sintering time at 1100°C led to an increase in the permittivity of CCTO. This is to be expected as many researchers believe that the giant permittivity is a result of CCTO being similar to an IBLC.[4,8] The effective dielectric constant for an IBLC is:[11]

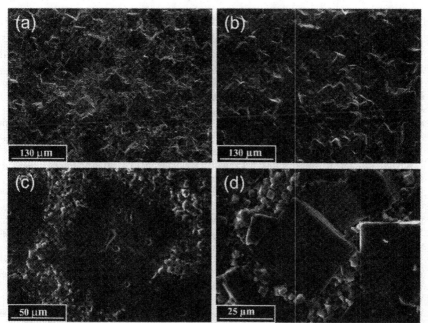

Fig. 3. SEM fractographs of the CANT powder sintered at 1100°C for 3 h (a &d) and 16 h (b) as compared to CSTD CCTO powder sintered at the same temperature for 3 h (c).

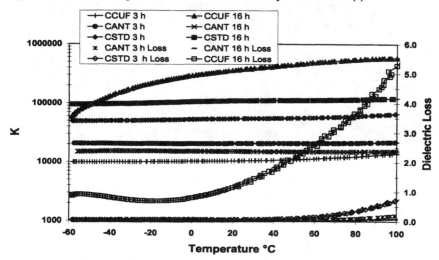

Fig.4 Dielectric properties (1 kHz) of the 3 powders sintered for 3 and 16 h at 1100°C.

$$k_{eff} = k_{gb}d_g/t \qquad (1)$$

where k_{gb} and t are the dielectric constant and thickness of the insulating grain boundary layer and d_g is the average grain size. Therefore with increasing grain size k_{eff} should increase. In the case of CCTO powder sintered at 1100°C for 2 to 8 h there was a bimodal distribution of large grains and small grains. To account for this type microstructure Pan et al.[10] have developed a model that derives k_{eff} as a function of both the volume fraction and size of the small and large grains. The model shows that with increasing volume fraction of large grains k_{eff} increases too. However, unlike the sintering time data of Fang et al.[12] for 1065°C, which shows a saturation in the dielectric constant with increasing sintering time after the microstructure has reached close to 100 volume percent of large grains, our dielectric constant data shows a continual increase in permittivity. Possible reasons for this dissimilarity are differences in grain size and/or resistivity. It is possible that with increasing sintering time at a higher temperature of 1100°C more substantial grain growth will occur than when compared to sintering at 1065°C. Also others researchers have shown that the permittivity of CCTO will increase with a decrease in the resistivity of CCTO. Li et al.[5] showed that as the sintering temperature of CCTO increased from 1000 to 1100°C its resistivity dropped while its permittivity increased. Similar results have been observed in systems such as PFN which exhibit dielectric behavior comparable to CCTO.[8] Therefore, at the higher sintering temperature of 1100°C it is possible more oxygen vacancies are being created with increasing sintering time and therefore the resistance of the CCTO grains is being lowered which leads to a higher dielectric constant.[5] One other possibility is that with increasing sintering time a different grain boundary phase is evolving as shown in Fig. 1(h). Energy dispersive analysis on unpolished surfaces indicated that these grain boundary phases are Cu rich. However, extensive TEM microanalysis is needed to confirm this possibility.

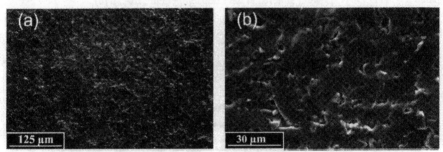

Fig. 5. SEM fractographs of CANT powder sintered at 1100°C for (a) 3 h and (b) 16 h.

The Effect of Anatase as a Precursor for CCTO on its Dielectric Properties

Anatase was chosen to replace rutile as the TiO_2 precursor for synthesizing CCTO. The reason for this choice is that anatase undergoes a reconstructive phase transformation at 915°C to form rutile.[13] This reaction is an exothermic reaction with a small volume change that may act as a driving force for improved sintering. Kim et al.[13] have shown this to be true as titania sintered from anatase reached 15% shrinkage at a temperature of 70°C below that for titania sintered from rutile. Porosity data indicates that this is a possibility for $CaCu_3Ti_4O_{12}$ also as the porosity dropped from 6 to 4% when sintering at 1100°C for 3h using the anatase-based powder, CANT. Improved sintering rates may also be responsible for the dramatic decrease in the presence of large pores in the large grains as contrasted in Figures 3(c) and (d). Typically, in abnormal grain growth grains coarsen faster than the material sinters resulting in the presence of entrapped pores. However, if the sintering rates are improved the pores can be annihilated in time before they are trapped leading to pore-free grains as observed in Figure 3(d). It is believed that this reduction in the porosity of the big grains is responsible for the dramatic decrease in dielectric loss as the magnitude of the loss tan can be very sensitive to small changes in porosity.[14] The loss dropped from 0.050 to 0.028 for samples sintered for 3 h and from 0.053 to 0.010 for 16 h.

The use of anatase also effected the grain-size distribution. Though both the CSTD and CANT powders sintered at 1100°C for 3 h show approximately the same volume percentage of small grains (27%) their grain size distributions are different. The typical large grain size range for the CANT sample was 5 to 60 microns in size (median grain size- 25microns) as compared to 30 to 160 microns for the CSTD material (median grain size- 67 microns).

This difference in grain-size distribution may be responsible for the drop in permittivity for the anatase-based material as compared to its rutile-based counterpart. The room temperature constant for CCTO dropped from 54,862 to 20,749. IBLC theory may explain this drop because the ratio of the drop (2.64) is very similar to the ratio in the difference in the median grain size of the two samples (67/25- 2.68). However, the theory can not explain why the dielectric properties of the CANT sample declined when sintered at longer times (see Fig. 4). It is unknown why the relative density of the sample declined to 94.3%. The increase in sintering time lead to a CCTO ceramic with outstanding properties- a room temperature dielectric loss of 0.010 and a dielectric constant of 15,000, which changed less than 4% over the temperature range of -50 to 100°C.

The Effect of Acid-Treated Powder on the Microstructure and Dielectric Properties of CCTO
The presence of a bimodal non-uniform grain size distribution with the presence of flat grains indicated abnormal grain growth was occurring.[15] This same phenomena was also observed for a CuO-doped BSTZ IBLC.[9] Therefore, it was felt that the presence of free CuO was responsible for the discontinuous grain growth observed in the CSTD samples. This was a real possibility because after three calcinations XRD revealed the presence of excess CuO in the starting powder even though no copper oxide was detected by XRD after sintering.[6] The CSTD powder was acid-treated with 3M HCl to remove this extra CuO. XRD did not detect any CuO after the treatment. This led to dramatic changes in the microstructure of the CCTO ceramic as shown in Figure 5. No substantial amounts of abnormal grain growth were observed in the CuO-free powder, CCUF, even after sintering for 16 h at 1100°C. This development of a small-grained $CaCu_3Ti_4O_{12}$ ceramic led to an 80% reduction in its room temperature permittivity as compared to the CSTD powder (see Fig. 4). The reduction in permittivity was expected by the IBLC model as the median grain size of the CSTD powder (67 microns) is substantially larger than the average grain size of the CCUF powder (1.7 microns).

The order of magnitude increase in permittivity for the CCUF powder sintered for 16 h (see Fig. 4) was unexpected. There was no substantial increase in grain size to account for this. The behavior of the dielectric constant versus temperature is also unexpected as the permittivity of the 16 h sample varies linearly with increasing temperature as opposed to being nearly constant for the 3 h sample. The sample is also very lossy. This lossiness could effect LCR meter measurements leading to an apparent higher permittivity. Research on this possible measurement artifact is ongoing.

107

CONCLUSION

CaCu$_3$Ti$_4$O$_{12}$ synthesized from standard precursors showed the presence of abnormal grain growth when sintered at 1100°C for 2 h resulting in large flat porous grains in a fine-grained matrix. With increasing sintering time the volume of fine grains diminished to zero when sintered for 12 h at 1100°C. The improvement in permittivity with increasing sintering time was correlated with increasing grain size and volume percent of large grains using the standard IBLC model. CaCu$_3$Ti$_4$O$_{12}$ synthesized using anatase to replace rutile as the titania precursor showed evidence of enhanced sintering. This was evidenced by the reduction of the large pores in the large grains of the CCTO ceramic. It also led to the reduction in median grain size of the material as compared to the standard powder sintered for 3 h at 1100°C. As a result of the overall reduction in average grain size its permittivity was reduced but with the reduction in porosity significant reduction in dielectric loss was measured. CaCu$_3$Ti$_4$O$_{12}$ synthesized with acid-treated powder showed no signs of abnormal grain growth. The reduction of free CuO was responsible for this behavior. This lead to the development of a fine-grained ceramic dielectric with reduced permittivity when sintered at 1100°C for 3h. Sintering at 1100°C for 16 h led to a fine-grained ceramic with a permittivity of over 350,000 which could not be explained the IBLC model. However, the material is very lossy which may lead to artificially high permittivity LCR meter measurements.

REFERENCES

[1]C. C. Homes, T. Vogt, S. M. Shapiro, S. Wakitomo, A. P. Ramirez, "Optical Response of High-Dielectric-Constant Perovskite-Related Oxide," *Science*, 293, 673-76 (2001).

[2]M. A. Subramanian, D. Li, N. Duan, B. A. Reisner, and A. W. Sleight, "High Dielectric Constant in ACu$_3$Ti$_4$O$_{12}$ and ACu$_3$Fe$_3$O$_{12}$ Phases," *J. of Solid State Chem.*, 151, 323-25 (2000).

[3]R. N. Choudhary and U. Bhunia, "Structural, Dielectric, and Electrical Properties of ACu$_3$Ti$_4$O$_{12}$ (A=Ca, Sr, and Ba)," *J. Mater. Sci.*, 37, 5177-82 (2002).

[4]T. B. Adams, D. C. Sinclair, and A. R. West, "Giant Barrier Layer Capacitance Effects in CaCu$_3$Ti$_4$O$_{12}$ Ceramics," *Adv. Mater.*, 14, 1321-23 (2002).

[5]J. Li, K. Cho, N. Wu and A. Ignatiev, "Correlation Between Dielectric Properties and Sintering Temperatures of Polycrystalline CaCu$_3$Ti$_4$O$_{12}$," *IEEE Trans*,1070-9878, 534-41 (2004).

[6]B. A. Bender and M.-J. Pan, "The Effect of Processing on the Giant Dielectric Properties of CaCu$_3$Ti$_4$O$_{12}$," *Mat. Sci. Eng. B.*, 117, 339-47 (2005).

[7]L. He, J. B. Neaton, D. Vanderbilt, and M. H. Cohen, "Lattice Dielectric Response of CdCu$_3$Ti$_4$O$_{12}$ and CaCu$_3$Ti$_4$O$_{12}$ From First Principles," *Phys. Rev. B*, 67, 012103-1-4 (2003).

[8]I. P. Raevski, S. A. Prosandeev, A. S. Bogatin, M. A. Malitskaya, and L. Jastrabik, "High Dielectric Permittivity in AFe$_{1/2}$B$_{1/2}$O$_3$ Nonferroelectric Perovskite Ceramics (A=Ba, Sr, Ca; B=Nb, Ta, Sb)," *J. Appl. Phys.*, 93, 4130-37 (2003).

[9]C.-F. Yang, "Improvement of the Sintering and Dielectric Characteristics of Surface Barrier Layer Capacitors by CuO Addition," *Jpn. J. Appl. Phys.*, 35, 1806-13 (1996).

[10]M.-J. Pan and B. A. Bender, "Dielectric Constant of CaCu$_3$Ti$_4$O$_{12}$ Ceramics with a Bimodal Grain Size Distribution," accepted for publication (3/05) in *J. Am. Ceram. Soc.*

[11]B.-S. Chiou, S.-T. Lin, J.-G. Duh, and P.-H. Chang, "Equivalent Circuit Model in Grain-Boundary Barrier Layer Capacitors," *J. Am. Ceram. Soc.*, 72, 1967-75 (1989).

[12]T.-T. Fang and H.-K. Shiau, "Mechanism for Developing the Boundary Barrier Layers of CaCu$_3$Ti$_4$O$_{12}$," *J. Am. Ceram. Soc.*, 87, 2072-79 (2004).

[13]D.-W. Kim, T.-G. Kim, and K. S. Hong, "Low-Firing of CuO-Doped Anatase," *Mat. Res. Bull.*, 34, 771-781 (1999).

[14]S. J. Penn, N. M. Alford, A. Templeton, X. Wang, M. Xu, M. Reece, and K. Schrapel, "Effect of Porosity and Grain Size on the Microwave Dielectric Properties of Sintered Alumina," *J. Am. Ceram. Soc.*, 80, 1885-88 (1997).

[15]B.-K. Lee, S.-Y. Chung, and S.-J. L. Kang, "Grain Boundary Faceting and Abnormal Grain Growth in BaTiO$_3$," *Acta. Mater.*, 48, 1575-1580 (2000).

DIELECTRIC AND MICROSTRUCTURAL PROPERTIES OF Ba(Ti₁₋ₓZrₓ)O₃ THIN FILMS ON COPPER SUBSTRATES

J. F. Ihlefeld and J-P. Maria
Department of Materials Science and Engineering
North Carolina State University
Raleigh, North Carolina 27606

W. Borland
DuPont Electronic Technologies
Research Triangle Park, North Carolina 27709

ABSTRACT

Barium titanate zirconate, $Ba(Ti_{1-x}Zr_x)O_3$ ($0 \leq x \leq 0.25$), thin films were deposited via the chemical solution deposition (CSD) process directly on copper foils. The films were processed in a reductive atmosphere containing water vapor and hydrogen gas at 900°C in order to preserve the metallic copper substrate while crystallizing the film into a perovskite structure. The microstructure and phase transition phenomena of films were studied utilizing x-ray diffraction, atomic force microscopy, and the temperature dependence of the dielectric constant and loss tangent. Increasing the fraction of $BaZrO_3$ revealed several effects, including an increase in unit cell dimensions, a decrease in both the temperature and value of the maximum permittivity, as well as a decrease in the average grain size of the films. Films were analyzed for dispersion in the transition temperature with frequency. Results indicated that films containing 25 mol% $BaZrO_3$ demonstrate a shift in the temperature of the ferroelectric phase transition with increasing measurement frequency. This shift, combined with the dispersive nature of the transition, suggests that films of this composition are of the relaxor ferroelectric family. The ability to process high permittivity thin film materials directly on inexpensive copper substrates has strong technological implications toward embedded passives and efficient frequency agile devices.

INTRODUCTION

Barium titanate zirconate (BTZ) is a well-known ferroelectric material that has garnered attention for applications where its tunable dielectric constant or high permittivity can be exploited.[1,2] The addition of barium zirconate to barium titanate allows for a variety of interesting phenomena to occur. Much like strontium titanate, barium zirconate has been used to shift the Curie point to lower temperatures to allow for the maximum in permittivity to occur in a temperature regime more favorable for room temperature applications. In contrast to the alloying of barium titanate with strontium titanate however, the addition of zirconium shifts the other phase transitions to higher temperatures resulting in a Curie temperature pinching effect.[3,4] This pinching effect increases and broadens the overall peak in permittivity and lowers the thermal coefficient of capacitance. The temperature response of BTZ has also been investigated and relaxor ferroelectric characteristics were identified. A diffuse phase transition and dispersion in the permittivity with frequency,[5] and dispersion in the permittivity with an applied DC bias have been reported.[6] It is accepted that in bulk ceramics, a composition containing 26 mol % zirconium shows relaxor behavior while those with less zirconium display a diffuse phase transition with no frequency dispersion.[7] Several groups have investigated the dielectric and relaxor properties of this material in a thin film embodiment.[1, 8-10] It has been suggested that the onset of relaxor behavior in thin films can be realized in compositions containing as low as 15 mol % zirconium.[9] Most of this thin film work however, has utilized depositions of films on refractory or noble metal substrates. While practical for many applications, these substrate materials can impart limitations on other uses where substrate conformality or cost may be a factor. Recently it has been demonstrated that ferroelectric thin films can be deposited on base-metal substrates. This has been accomplished in one of 3 ways: (1) the exploitation of thermodynamics to process materials in regimes where film cations are oxidized and crystallized into the desired perovskite structure and substrate materials are reduced,[11-14] (2) through sophisticated binder removal methods with a chemical solution deposition (CSD) process,[15] or (3) by utilizing interfacial phases between the crystallized film and substrate.[16,17] The use of ferroelectric films on base metals has been demonstrated for use in coated conductors,[18] frequency agile devices,[14] and as decoupling capacitors.[13] This technology is of particular interest to the embedded passive community where the low cost, high conductivity, and flexible substrates are ideal for facilitating the replacement of surface mount components with passive devices embedded into the layers of a printed wiring board. The present work focuses on advancing the knowledge base of complex oxide films processed on base metal substrates by demonstrating the deposition of barium titanate zirconate thin films directly on copper foil substrates via CSD methods. In doing so, the microstructural, crystallographic, and dielectric properties of this material are investigated, including the existence of relaxor behavior and the effect of varying levels of zirconium on the phase transition phenomena.

EXPERIMENTAL

Chemical solutions were prepared with compositions containing Ti/Zr ratios of 95/5, 90/10, and 75/25. Solutions were formed through a modified chelate method by reacting titanium isopropoxide with zirconium propoxide (70 weight % in propanol) in the prescribed ratios and further chelating of the titanium component with acetylacetone and diethanolamine. The titanium/zirconium precursor solutions were then combined with barium acetate dissolved in glacial acetic acid such that the A:B site ratio was 1:1. The solutions were diluted to 0.3 molar with the addition of methanol. Stoichiometry of the solutions was determined and controlled by masses of the constituents.

Solutions were spin-casted on untreated Oak-Mitsui 18 micron thick PLSP copper foils at 3000 RPM for 30 seconds. Films were dried on a hotplate at 250 °C for 7 minutes and 30 seconds. This process was repeated 3 times to achieve the desired film thickness of 600 nm. The films were fired in a reducing atmosphere of reagent grade nitrogen, hydrogen containing forming gas, and water vapor at 900 °C for 30 minutes. The equilibrium reaction of water vapor, hydrogen, and residual oxygen in the nitrogen carrier gas is sufficient to provide for atmospheres amenable to oxidizing the cation components in the film while reducing and preserving the base-metal copper substrate. Total furnace pressure during the crystallization anneal was 1 atm with a partial pressure of oxygen of 10^{-11} atm as measured in situ by a solid state oxygen sensor. The films were reoxidized under vacuum at 10^{-6} atm of oxygen and 550°C for 30 minutes to reduce the concentration of oxygen point defects associated with the high-temperature and low partial pressure of oxygen present during the crystallization anneal. Platinum top electrodes $\sim 10^{-4}$ cm^2 and 50 nm thick were applied via RF magnetron sputtering through a shadow mask to define the metal-insulator-metal capacitor structures.

Phase development, crystallinity, and lattice parameters of the films were investigated with a Bruker AXS D-5000 x-ray diffractometer equipped with a GADDS area detector. Lattice parameters were calculated utilizing a Nelson-Riley correction function and assuming a pseudocubic crystal symmetry for all compositions. A CP Research Thermomicroscope atomic force microscope was used to characterize film surface morphology and grain size. Film thickness was determined with transmission electron microscopy (TEM). Capacitance and loss tangent data was collected with an HP 4192A Impedance Analyzer and a MMR Technologies Inc. cryogenic temperature stage. Temperature response of the dielectric properties of each film were obtained from −175 to 225 °C and at frequencies of 1, 10, and 100 kHz with an oscillator level of 0.05 volts to examine relaxor behavior.

RESULTS AND DISCUSSION

Figure 1 shows the x-ray diffraction patterns for the as-crystallized barium titanate zirconate thin films on copper foils. The patterns display narrow peak widths with low and flat background levels, consistent with barium titanate thin films deposited on copper substrates.[13] Further characterization of the *001* family of peaks, with the aid of the Nelson-Riley function for correction of systematic diffractometer errors,[20] revealed the dependency of lattice parameter with the mole fraction of $BaZrO_3$ as shown in Figure 2. As expected, the substitution of zirconium for titanium results in an enlarged cell dimensions. The accepted values for lattice parameter for $BaZrO_3$ is 4.183 angstroms and pseudo-cubic $BaTiO_3$ has a lattice dimension of 4.008 angstroms, based on cell volume for a bulk ceramic material.[19] The increase in cell dimensions is approximately linear with respect to mole fraction of substituted zirconium. This effect is indicative of a continuous increase in cell volume as zirconium dopant fraction increases.

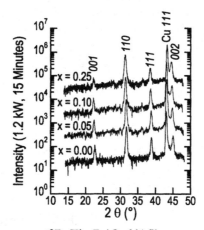

Figure 1. θ - 2θ x-ray diffraction scans of $Ba(Ti_{1-x}Zr_x)O_3$ thin films on copper foil.

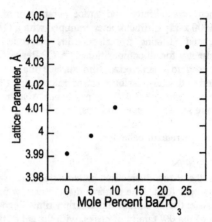

Figure 2. Lattice parameters of Ba(Ti$_{1-x}$Zr$_x$)O$_3$ thin films deposited on copper foils, as extrapolated using the *001* reflections and a Nelson-Riley correction function.

The dielectric constant versus temperature response curves are compared at 10 kHz for each composition in Figure 3. It is established that increasing the mole fraction of barium zirconate results in a shift of the relative permittivity maximum to lower values as well as lower temperatures. Figure 4 plots the temperature and value of the dielectric maximum versus the composition. It is observed that the temperature of the ferroelectric anomaly decreases linearly with mole percent BaZrO$_3$. This trend is slightly more exaggerated than that observed by Verbitskaia et al.[4] in bulk ceramic material, however the effect of fine grain size to broaden the transition behavior must be considered.

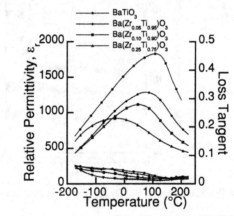

Figure 3. Dielectric constant versus temperature for Ba(Ti$_{1-x}$Zr$_x$)O$_3$ thin films deposited on copper foil substrates.

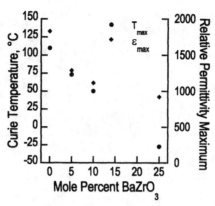

Figure 4. Curie temperature and maximum permittivity values plotted versus composition for Ba(Ti$_{1-x}$Zr$_x$)O$_3$ thin films deposited on copper foil substrates.

Fig. 5 shows the error signal atomic force microscope scans taken in contact mode of the 4 film compositions; increasing the fraction of BaZrO$_3$ results in a decrease in average grain size. The decrease in grain size with increase in fraction of substituent is perhaps due to the hindrance in diffusion provided by the additional, and highly refractory, Zr additions. Alternatively, the addition of Zr may increase the nucleation rate of BTZ crystals, thereby limiting the final grain size. Since a reduction in the average crystallite size in perovskite ferroelectrics can lead to a suppressed Curie temperature,[21,22] the more drastic decrease in transition temperature for barium titanate zirconate thin films deposited on copper substrates in comparison with bulk ceramics of similar compositions can be understood. The limited thermal budget of thin film processing versus bulk ceramic processing does not allow for equality of grain sizes for the various composition films.

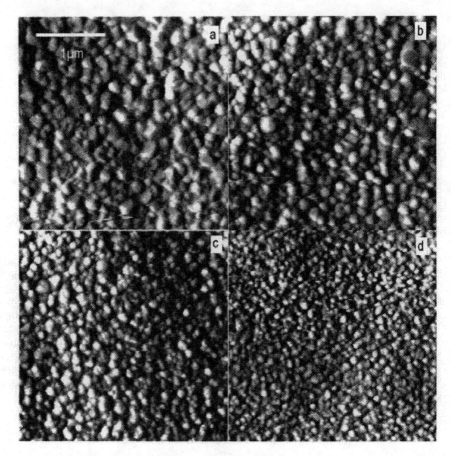

Figure 5. 3 μm x 3 μm contact mode AFM error signal scans of a) BaTiO₃ b) Ba(Ti$_{0.95}$Zr$_{0.05}$)O₃ c) Ba(Ti$_{0.90}$Zr$_{0.10}$)O₃ and d) Ba(Ti$_{0.75}$Zr$_{0.25}$)O₃ on copper foil substrates.

Dielectric relaxation was studied by observing the affect of oscillator frequency on the temperature of the Curie point for each composition. As shown in Figure 6, samples with compositions of 10 mol% BaZrO₃ or less do not show any dispersion in the Curie temperature. However, in the film of a composition containing 25 mol% BaZrO₃ the dielectric maximum appears to shift as measurement frequency is increased. This, combined with the diffuse nature of the phase transition, suggests that the Ba(Ti$_{0.75}$Zr$_{0.25}$)O₃ film exhibits relaxor behavior. This data is consistent with that measured by other groups on thin films of barium titanate zirconate deposited on refractory substrates.[9]

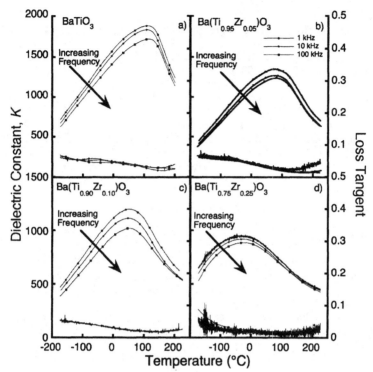

Figure 6. Temperature and frequency dependency of dielectric constant and loss tangent for a) BaTiO$_3$ b) Ba(Ti$_{0.95}$Zr$_{0.05}$)O$_3$ c) Ba(Ti$_{0.90}$Zr$_{0.10}$)O$_3$ and d) Ba(Ti$_{0.75}$Zr$_{0.25}$)O$_3$ on copper foil substrates. The arrows indicate curves measured at increasing frequencies from 1kHz to 100 kHz.

CONCLUSIONS

Barium titanate zirconate thin films were deposited directly on copper foil substrates via a chemical solution deposition technique. The films were crystallized in a reductive atmosphere, such that the copper substrate is preserved and the perovskite phase of the deposited film forms. The films were reoxidized in a slightly more oxidizing atmosphere to minimize electrically active point defects resulting from low partial pressure of oxygen processing. X-ray diffraction measurements reveal that as the fraction of barium zirconate is increased, the lattice parameter of the material increases linearly. Permittivity versus temperature plots reveal that increasing levels of barium zirconate decrease the temperature of the ferroelectric anomaly and value of the permittivity maximum. AFM topographical scans show that the increased levels of barium zirconate decrease the average grain size in the films. Relaxor ferroelectric behavior was indicated in a barium zirconate titanate thin film with a composition containing 25 mol% barium zirconate as demonstrated by an apparent shift in the temperature of the ferroelectric anomaly with increasing measurement frequency.

REFERENCES

[1]Hofer, C.; Hoffmann, M.; Boettger, U.; and Waser, R., *Ferroelectrics,* **270,** 179-184 (2002).

[2]Mizuno, Y.; Morita, K.; Hagiwara, T.; Kishi, H.; Ohnuma, K.; and Ohsato, H., *Jpn. J. Appl. Phys.,* **43** (9B), 6640-6644 (2004).

[3]Jonker, G.H., *Philips Technical Review,* **17** (5), 129-137 (1955).

[4]Verbitskaia, T.N.; Zhdanov, G.S.; Venevtsev, I.N.; and Soloviev, S.P., *Soviet Physics. Crystallography,* **3,** 182-192 (1958).

[5]Cross, L. E., *Ferroelectrics,* **151,** 302-320 (1994).

[6]Wada, S.; Adachi, H.; Kakemoto, H.; Chazono, H.; Mizuno, Y.; Kishi, H.; and Tsurumi, T., *J. Mater. Res.,* **17** (2), 456-464 (2002).

[7]Ravez, J.; and Simon, A., *Eur. J. Solid State Inorg. Chem.,* **34,** 1199-1209 (1997).

[8]Dixit, A.; Majumder, S.B.; Dobal, P.S.; Katiyar, R.S.; Bhalla, A.S., *Thin Solid Films,* **447-448,** 284-288 (2004).

[9]Tohma, T.; Masumoto, H.; and Goto, T., *Jpn. J. Appl. Phys.,* **42** (Part 1, 11), 6969-6972 (2003).

[10]Tanaka, K.; Suzuki, K.; Fu, D.; Nishizawa, K.; Miki, T.; and Kato, K., *Jpn. J. Appl. Phys.,* **43** (9B), 6525-6529 (2004).

[11]Dawley, J.T.; Ong, R.J.; and Clem, P.G., *J. Mater. Res.,* **17** (7), 1678-1685 (2002).

[12]Ong, R.J.; Dawley, J.T.; and Clem, P.G., *J. Mater. Res.,* **18,** 2310 (2003).

[13]Ihlefeld, J.; Laughlin, B.; Hunt-Lowery, A.; Borland, W.; Kingon, A.; and Maria, J-P., *J. Electroceram.,* **14** (2), 95-102 (2005).

[14]Laughlin, B.; Ihlefeld, J.; and Maria, J-P., *J. Amer. Cer. Soc.,* In Publication (2005).

[15]Losego, M.D.; Jimison, L.H.; Ihlefeld, J.F.; and Maria, J-P., *Appl. Phys. Lett.,* Under review (2005).

[16]Maria, J-P.; Cheek, K.; Streiffer, S.; Kim, S-H.; and Kingon, A., *J. Amer. Cer. Soc,* **84** (10), 2436-2438 (2001).

[17]Kim, T; Kingon, A.I.; Maria, J-P.; and Croswell, R.T., *J. Mater. Res.,* **19** (10), 2841-2848 (2004).

[18]Siegal, M.P.; Clem, P.G.; Dawley, J.T.; Ong, R.J.; Rodriguez, M.A.; and Overmyer, D.L., *Appl. Phys. Lett.,* **80,** 2710 (2002).

[19]International Centre for Diffraction Data, Powder Diffraction Data, **31,** (1988).

[20]Cullity, B.D.; and Stock, S.R., in Elements of X-Ray Diffraction (Prentice Hall, Upper Saddle River, New Jersey, 3rd Edition, 2001) p. 366.

[21]Arlt, G.; Hennings, D.; and de With, G., *J. Appl. Phys.* **58** (4), 1619-1625 (1985).

[22]Parker, C.B.; Maria, J-P.; and Kingon, A.I., *Appl. Phys. Lett.* **81** (2), 340-342 (2002).

EFFECT OF A-SITE SUBSTITUTIONS ON THE MICROSTRUCTURE AND DIELECTRIC PROPERTIES OF BISMUTH SODIUM TITANATE-BASED CERAMICS EXHIBITING MORPHOTROPIC PHASE BOUNDARY

R. Vintila, G. Mendoza-Suarez, J. A. Kozinski, R. A. L. Drew
McGill University, 3610 University St. Montreal, Quebec, H3A2B2

ABSTRACT

This study reports on the microstructure and dielectric properties of calcium and strontium solid solutions in bismuth-sodium-titanate (BNT) ceramics with composition $Ca_x(Bi_{0.5}Na_{0.5})_{1-x}TiO_3$ and $Sr_y(Bi_{0.5}Na_{0.5})_{1-y}TiO_3$, which exhibit morphotropic phase boundaries. The results showed, that the best dielectric properties in the high temperature region were obtained for x=0.05 and y= 0.27 near room temperature. All the ceramic compositions exhibited a single-phase perovskite structure. Strontium additions lowered both the antiferroelectric phase transition temperature and Curie point. Further, substituting bismuth with lanthanum on the A site in both BNT-CT and BNT-ST systems yielded lower smaller grain size and lower dielectric loss, of BNT-based ceramics.

INTRODUCTION

Hydrogen can be produced from natural gas in plasmas generated in dielectric barrier discharge (DBD) reactors. Dielectric materials such as ceramics, glasses or polymers are placed between two electrodes to produce microdischarges, which are intended to promote cracking of CH_4 molecules. The most suitable ceramics are those showing moderately high dielectric constants (~3000) at relatively elevated temperature (room temperature to 300°C), along with reasonably broad permittivity curves and low dielectric losses. High dielectric breakdown strength is also of prime importance.

Most of the dielectric materials in current use as ceramic capacitors are based on barium titanate, $BaTiO_3$. However, beyond 125°C, the permittivity of this class of dielectrics falls to very small values, making them unsuitable for the DBD process. In order to overcome this challenge, the dielectric properties of a new family of materials based on bismuth sodium titanate $Bi_{0.5}Na_{0.5}TiO_3$ (BNT) were investigated. These compounds have a perovskite (ABO_3) structure with rhombohedral symmetry between room temperature and 200°C, showing a paraelectric phase around 330°C[1]. Compositions, like BNT having an antiferroelectric phase over a wide temperature region will manifest temperature independent high permittivity and improved field stability. This composition works very well above room temperature and benefits from high dielectric constant over a wide temperature range.

A noticeable feature of ferroelectric materials is the occurrence in the phase diagram of a morphotropic phase boundary (MPB), which separates tetragonal and rhombohedral ferroelectric regions. Solid solutions with compositions close to the MPB show the best electromechanical properties[2], and will manifest high mechanical strength and remarkably improved dielectric characteristics compared with nonmodified BNT ceramics[3]. From this point of view the study was based on BNT- solid solutions, which fulfill the MPB criteria and therefore, single and two cation substitutions in the A-site (Na and Bi sites) of the perovskite structure were performed.

Predictions by Takenaka et al.[4] showed that the amount of A-site (mol%) ions needed to form the MPB decreases as function of the ionic radii of the substituting ions. In other words, the larger the ionic radii for the A-site ions, the smaller the amount of the substitution needed to form the MPB. Therefore, the substitution of a large divalent ion (e.g. Ca) for bismuth and sodium on the A-site $[Ca_x(Bi_{0.5}Na_{0.5})_{1-x}TiO_3]$ will lead MPB for very small amounts of Ca^{2+}, in the compositional range $0.01 < x < 0.1$. Furthermore, according to Sakabe et al.[5], Ca-doped $BaTiO_3$ presented high dielectric constant and stable temperature dependence as well as a fine, more uniform microstructure than the undoped $BaTiO_3$ for very small doping levels. Therefore, due to the similarities of BNT and $BaTiO_3$ the BNT-CT solid solution could be a promising system for MPB.

For smaller ions such as Sr, the amount of substitution should be higher. According to Cho et al[6], MPB in the $Sr_y(Bi_{0.5}Na_{0.5})_{1-y}TiO_3$ should be achieved in the $0.25 < y < 0.26$ compositional range. For Sr substitution, both transition temperatures (ferroelectric to antiferroelectric originally at 220°C and antiferroelectric to paraelectric at 320°C in BNT shift to lower temperature and the relative dielectric constant increases[7]. At the microstructural side, dielectric properties and especially dielectric constant depend on microstructure and fine and uniform grain size lead to enhanced dielectric and electromechanical properties. A few types of dopants are known to affect the microstructure characteristics as well as the electrical properties. This task could be achieved by small amounts of lanthanum (III) oxide added to BNT, which leads to a smaller grain size and a more uniform grain size distribution[8]. Therefore, a cooperative soft doping effect of La and Ca for the BNT-CT system, and analogously, La and Sr for the BNT-ST system, could enhance the dielectric and microstructural properties of the ceramics under investigation.

EXPERIMENTAL

Ceramic samples were prepared employing the conventional ceramic sintering technique. The samples were prepared using of Bi_2O_3, Na_2CO_3, TiO_2, CaO, SrO and La_2O_3 99.99%+ purity from Ferro Corporation. The investigated compounds are $(Na_{1/2}Bi_{1/2})_{1-x}Ca_xTiO_3$, where $x = 0.01$, 0.05, 0.1 and $(Na_{1/2}Bi_{1/2})_{1-y}Sr_yTiO_3$ where $y = 0.26$, 0.27, 0.30 (Table I).

Table I – BNT compositional modifications

Solid	$Bi_{0.5}Na_{0.5}TiO_3$-$xCaTiO_3$			$Bi_{0.5}Na_{0.5}TiO_3$ –$ySrTiO_3$		
Solutions	BNT-1CT	BNT-5CT	BNT-10CT	BNT-26ST	BNT-27ST	BNT-30ST

For comparison reasons, compositions with formulae BNT-50CT and BNT-50ST were also fabricated. Furthermore in the attempt to improve BNT ceramic properties, lanthanum doped ceramics: $Na_{0.5}(La_{0.1}Bi_{0.4})TiO_3$-1CT (BNTL-1CT) and $Na_{0.5}(La_{0.1}Bi_{0.4})TiO_3$-26ST (BNTL-26ST) were investigated.

The powders were batched in adequate stoichiometric proportions and then ball milled for 10 hours in water based systems. The milled powders were then pressed into cylindrical disks of 25.4mm diameter and 10 mm thickness. Calcination was conducted at 800-900°C for 2-6 hours depending on the composition. The pre-sintered disks were ground to powder form. The powder was used for phase identification by XRD. The pre-sintered powder was again ball-milled in water for 10 hours into fine powder. The dried cake was then mixed thoroughly with a PVA binder solution. The slurry was then spray dried to obtain a uniform flowing powder and then uniaxially pressed into disks of 25-40 mm diameter and different thickness at 50 – 150

118

MPa. Sintering tests, phase analyses and density measurements were conducted to determine the optimum sintering profiles for the various compositions. The final sintering stage of the pellets was done at 1150°C for 1 hour in air in a closed alumina crucible. Sintered powders of all compositions were prepared for XRD analysis. The microstructure of the sintered samples was determined from polished and chemically etched samples, using SEM. The dielectric measurements were performed at frequencies from 1kHz to 1MHz on silver electroded disks using a high precision LCR meter (HP 4824A). The temperature dependence of dielectric properties was measured in the temperature range of interest by placing the ceramics in an environmental chamber and monitoring the temperature with a thermocouple.

RESULTS AND DISSCUTION

A comparison between the X-ray patterns of the calcined and sintered BNT powders indicates that after sintering a single phase compound was formed. For the undoped BNT at the calcination temperature the material showed a mixture of phases that were identified to be $Bi_4Ti_3O_{12}$ and TiO_2, which completely disappear after the sintering cycle.

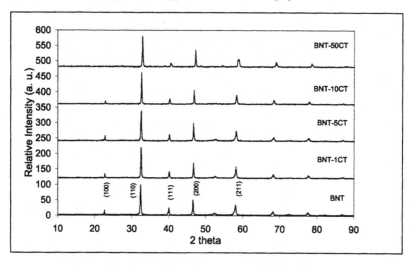

Figure1 XRD diffraction patterns for BNT-xCT ceramics

The XRD results indicate that the perovskite phase is formed after calcinations at 800°C for 2h, therefore no additional calcination was necessary. Figure 1 shows structure of the diffraction patterns of the synthesized BNT-xCT ceramics. The patterns indicate that the ceramics corresponds to single-phase perovskite structure. For the BNT-xCT (x=0-10) there are unresolved peaks of calcium in the Ca doped BNT indicating that calcium has gone into the lattice. The indices (hkl) of the reflecting planes were (100), (110), (111), (200), (211) as illustrated in Figure 1. The crystal structure was verified to be rhombohedral, similar to BNT at room temperature. Increase in calcium substitution introduces strains into the lattice and XRD patterns show small distortions as seen in Figure 2. As the calcium content increases (from BNT-

119

10CT to BNT-50CT) a second phase appears as demonstrated by the splitting of (100) or (111) peaks. Figure 2 shows a change in the diffraction peaks of Ca-doped BNT in the 38-43° 2θ region. Thus it could be concluded that in the case of doping Ca in BNT ceramics, Ca occupies the A site and hence will exhibit a single phase perovskite structure provided that the doping level is not too high (<10mol%), therefore the MPB for this system at room temperature could be placed between 5 and 10mol%.

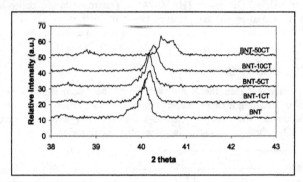

Figure 2 XRD pattern for BNT-xCT at 2θ between 38° and 43°

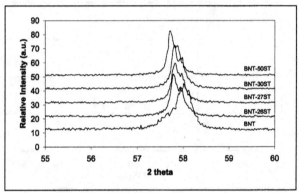

Figure 3 XRD pattern for BNT-yST at 2θ between 55° and 60°

For the BNT-yST system the phase development was also investigated. $SrTiO_3$ presents an ideal perovskite cubic structure and BNT has rhombohedral symmetry at room temperature; therefore the MPB exists at room temperature between the rhombohedral and cubic phases. When $SrTiO_3$ forms solid solutions with other perovskites it introduces smaller distortions in the lattice (Figure 3). The rhombohedral split disappears as the Sr content is increased and a sharp peak indicating cubic symmetry appears at compositions where y>30. Therefore for compositions higher than 30mol% ST the materials adopt cubic symmetry, so the MPB for this system at room temperature exists between 27 and 30mol%.

In the co-doped systems $Na_{0.5}(La_{0.1}Bi_{0.4})TiO_3$-1CT (BNTL-1CT) and $Na_{0.5}(La_{0.1}Bi_{0.4})TiO_3$-26ST (BNTL-26ST), due to the fact that the lanthanum content was very small, it was found that all the reflections corresponded to the perovskite with a rhombohedral structure, as the peak splitting indicative of the rhombohedral distortion is present in all samples. Therefore lanthanum incorporation into these compositions did not change the crystallographic structure.

The microstructure of the sintered samples was observed on polished, chemically etched cross sections of the samples using SEM. Figure 4 shows the micrographs of the BNT and BNT-xCT systems.

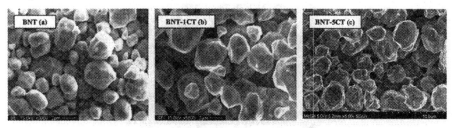

Figure 4 SEM Micrographs of BNT (a); BNT-1CT (b); BNT-5 (c)

As the calcium content increased the grain size slightly decreased and the samples showed a more uniform distribution as can be seen in Figure 4(c). In the unsubstituted BNT (Figure 4(a)), the grain size varied widely from 1-6µm, whereas in BNT-5CT the average grain size varied from 2-4µm. According to Saradhi et al.[9] calcium incorporation into the lattice not only increases the defects but also favors the formation of defect impurity complexes. Such complexes may also be formed in grain boundaries, which increase the grain boundary movement resistance and therefore reduces the grain growth.

Figure 5 Micrographs of BNT-1CT (a) and BNTL-1CT (b) ceramics

At the same time cation vacancies in La-substituted BNT could also segregate at the grain boundaries. Based on Yi et al.[10] theory, if the concentration of cation vacancies is higher than in the grains, positively charged La ions would be attracted by the cation vacancies and also segregate to the grain boundaries. Under these conditions the mobility of grain boundaries will be reduced, which will induce a uniform microstructure with smaller grains in the substituted ceramics as it could be seen in Figure 5 for BNT and BNTL-1CT.

The same effect was observed for the BNT-yST system. The grain size became smaller with the increase of strontium content and a more uniform size distribution was observed in the

La, Sr co-doped samples (Figure6). The average grain size for the BNTL-26ST was of the order of 1-3μm.

The dielectric constant and loss factor of the BNT-xCT ceramics at room temperature are shown in Figure 7(a). The dielectric constant values increased with the Ca content in the low frequency range and saturated at x = 10. For the BNT-yST ceramics (Figure 7(b)) the dielectric constant increased with the amount of Sr doping and saturated at y = 27.

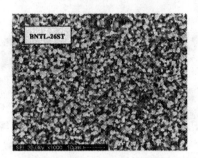

Figure 6 SEM micrograph of BNTL-26ST

The dielectric constant showed smaller values in the high frequency range at room temperature. An increase of frequency resulted also in a decrease of dielectric loss. In the MHz region there is no appreciable variation of dielectric loss in all samples and an increase in the amount of substituting atoms resulted in a flat response of the dielectric loss.

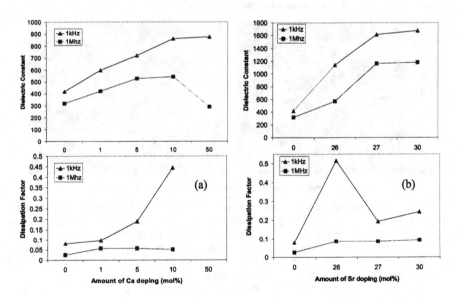

Figure 7 Dielectric constant and loss at R.T. as function of the amount of Ca (a) and Sr(b)doping

The values of dissipation factor are found to increase with calcium but decrease with frequency. In the kHz region an appreciable increase in dielectric loss as the calcium content is increased is observed, especially for x>10. The structural disorders and compositional fluctuations in the samples could be responsible for the observed behaviour.

For the BNT-yST, the dielectric constant value at room temperature increases with composition and reaches a plateau at y=27-30 (Figure 7(b)). However, there is an exceptional abnormal behaviour in the loss factor value for BNT-26ST ceramics.

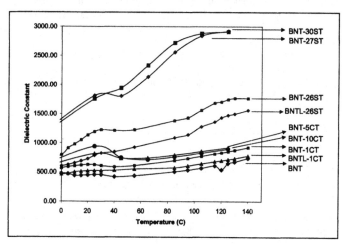

Figure 8 Temperature dependence of dielectric constant for the BNT and BNT-doped ceramics

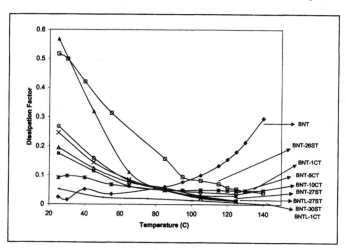

Figure 9 Temperature dependence of the loss tangent of the BNT and BNT-doped ceramics

For high temperature regions, above 80°C, all the BNT and substituted BNT samples show an increase with dielectric constant (Figure 8). The dissipation factor (Figure 9) of BNT-doped ceramics decreases with temperature. Interesting fact about the co-doped BNTL-1CT is that its loss tangent is almost constant through a wide temperature range ant to very small values.

CONCLUSIONS

From XRD pattern analyses the MPB occurs for BNT-CT at approximately 5-mol% Ca and for BNT-ST at approximately 27mol%Sr. La incorporation into the samples decreased the grain size and the loss factor for both BNT-CT and BNT-ST systems. The room temperature dielectric constant increased with the amount of doped ions reaching a plateau at compositions close to the MPB.

The dielectric constants of all the modified BNT compositions increase with temperature and are greater than that for the unsubstituted BNT. Moreover, an interesting fact about their loss factors is that loss factors of ceramics of all BNT compositions decrease with temperature except for those of the undoped BNT ceramic.

REFERENCES

[1] G.A. Smolenskii, V.A. Isupov, A.I. Agranovskaya, N.N. Krainik "New Ferroelectrics of Complex Composition IV" *Soviet Physics – Solid State*, **2**, 2651-54 (1961)

[2] T.Takenaka, K.Sakata, K.Toda, "Piezoelectric Properties of $(Bi_{1/2}Na_{1/2})TiO_3$ – Based ceramics", *Ferroelectrics* **106**, 375-380, (1990)

[3] J.-R. Gomah-Pettry, S.Said, P.Marchet, J.-P. Mercurio, "Sodium-bismuth titanate based lead-free ferroelectric materials", *Journal of the European Ceramic Society*, In Press

[4] T.Takenaka, A. Hozumi, T. Hata, K. Sakata, "Mechanical Properties of $(Bi_{1/2}Na_{1/2})TiO_3$ - based Piezoelectric Ceramics" *Silicates Industries*, **58** 136-42 (1993)

[5] Y. Sakabe, N. Wada, T. Hiramatsu, T. Tonogaki, "Dielectric Properties of Fined - Grained $BaTiO_3$ Ceramics Doped with CaO", *Jpn. J. Appl. Phys.*, **41**, 6922-6925 (2002)

[6] S. Y. Cho, S. E. Park, K. S. Hong, , "The variation of phase transition behavior on substituting Pb2+ and Sr2+ for A site cations in $(Na_{1/2} Bi_{1/2})TiO_3$ system"., *Ferroelectrics*, **195**, 27-30, (1997)

[7] K. Sakata, Y. Masuda, "Ferroelectric and Antiferroelectric Properties of $(Na_{0.5}Bi_{0.5})TiO_3$-$SrTiO_3$ Solid Solution Ceramics", *Ferroelectrics*, **7**, 347-349, (1974)

[8] H.-D. Li, C-D. Feng, W.-L. Yao, "Some effects of different additives on dielectric and piezoelectric properties of $(Bi_{1/2}Na_{1/2})TiO_3$-$BaTiO_3$ morphotropic phase boundary composition", *Materials Letters*, **58**, 1194-1198, (2004)

[9] B.V. B. Saradhi, K. Srinivas, T. Bhimasankaram, "Impedance and Modulus Spectroscopy of $(Na_{1/2}Bi_{1/2})_{1-x}Ca_xTiO_3$ Ceramics", *International Journal of Modern Physics B*, **16**, 4755-4766, (2002)

[10] J.Y.Yi, J.-K.Lee, K.-S. Hong, "Dependence of the Microstructure and the Electrical Properties of Lanthanum –Substituted $(Na_{1/2}Bi_{1/2})$ TiO_3 on Cation Vacancies", *J.Am. Ceram. Soc.* , **85**, 3004-3010, (2002)

HIGH Q (Ba, Sr)TiO₃ INTERDIGITATED CAPACITORS FABRICATED ON LOW COST POLYCRYSTALLINE ALUMINA SUBSTRATES WITH COPPER METALLIZATION

Dipankar Ghosh[1], B.Laughlin[1], J. Nath[2], A. I. Kingon[1], M. B. Steer[2] and J-P. Maria[1]

[1]The Electroceramic Thin Film Group, Department of Materials Science and Engineering, North Carolina State University, Raleigh, NC 27695,
[2]Electrical and Computer Engineering Department, North Carolina State University, Raleigh, NC 27695

ABSTRACT

Barium Strontium Titanate (BST) ferroelectric thin films are attractive for radio frequency and microwave applications. However, for many non-military uses, the high cost of conventionally processed devices is a limiting factor. This high cost stems from the use of single-crystalline sapphire, MgO, or LaAlO₃ substrates and Pt or Au metallization commonly used. Here we present a device process and materials complement offering a low cost alternative.

Planar interdigitated capacitors Ba₀.₇₅Sr₀.₂₅TiO₃ (BST) thin films with chromium/copper top electrodes were fabricated on polycrystalline alumina substrates using a single step photolithographic technique and lift-off. RF magnetron sputtering was used for fabrication of BST thin films while Cu thin films were thermally evaporated

The dielectric tunability of the Ba₀.₇₅Sr₀.₂₅TiO₃ IDCs was 40 % for an applied electric field of 120 kV/cm, which corresponds to 3 μm electrode gap spacing and a 35 volt dc bias. Low frequency (1MHz) loss measurements reveal a dielectric $Q \sim 100$ while a device Q of ~ 30 is obtained at 26 GHz. The reduction of Q between 0.1 and 26 GHz can be attributed to the metallization. Leakage current measurements of the BST planar varactors show current densities of 1.0×10^{-6} A / cm² for an electric field of 100 kV/cm. These dielectric characteristics (tunability and Q value) are comparable to numerous reports of IDCs with BST films prepared on expensive single crystalline substrates using noble metallization. As such, this technology is significantly less expensive, and amenable to large volume manufacturing.

INTRODUCTION

Currently there is a huge research interest in utilizing ferroelectric thin films for tunable microwave devices such as phase shifters, filters and matching networks[1-3]. Barium strontium titanate, Ba₁₋ₓSrₓTiO₃, $0 \le x \le 1$, (BST), a solid solution perovskite, is the material which has been widely investigated so far for microwave devices since it has high tunability, low loss and thermal stability when compared to other functional thin film oxides [2, 4]

Various techniques that have been used so far for BST thin film fabrication are metal organic chemical vapor deposition (MOCVD)[2,4], pulsed laser deposition (PLD)[5-7], RF sputtering[1,3] and sol gel processing[8] . Till date most of the BST thin film based microwave devices have been fabricated on expensive single crystal substrates such as MgO[5,9,10] , LaO[7,9,10], and Al₂O₃ (sapphire)[1,11] since high quality epitaxial BST thin films can be grown on such substrates at high temperature. However these are high cost substrates and are commercially available only in small size pieces. In this paper we have chosen polycrystalline alumina (Al₂O₃) substrate. These are low cost and available in bigger dimension which is suitable for large area film deposition. Also the size of the alumina substrates can be tailored to meet specific device requirements. Alumina also has attractive microwave properties such as low loss tangents (tan $\delta = 10^{-4}$ at 1 MHz) and a coefficient of thermal expansion (CTE ~ 9 ppm @ RT) which is very close to that of BST.

Traditionally the electrodes of choice for thin film oxide based devices are usually high cost noble metals such as Au, Pt and Ir[1, 2, 9] since they make good non-reactive electrodes. However due to the high resistance values of Pt and Au, multiple micron layer electrode thicknesses are necessary for achieving acceptable low sheet resistance of the device. This is expensive from a manufacturing point of view and also involves difficult patterning issues. Recently Cu has been introduced in the semiconductor integrated chip (IC) industry for use in interconnect lines. However limited work has been reported so far using it as an electrode material for thin film oxide based devices due to its inherently poor adhesive property and its tendency to oxidize[12]. Cu was chosen as the top electrode metal in our work since it provides the highest conductivity of any non noble metal and is also very low cost.

In this work we have fabricated BST thin films by radio frequency magnetron sputtering using only inexpensive substrate and metallization materials. The focus of this study is to demonstrate a single step, low cost, commercially viable technology for fabrication of BST based microwave devices.

EXPERIMENTAL PROCEDURE

In this paper we have used radio frequency magnetron sputtering technique to deposit $(Ba_{0.75}Sr_{0.25})TiO_3$ on 625 μm thick polished polycrystalline alumina substrates (Intertec Southwest Inc., Tucson, AZ). Sputtering was done from a 4" stoichiometric ceramic BST target. Sputtering was done off axis at an angle of 30° in an argon / oxygen gas mixture (Ar: $O_2 = 5:1$) at two different deposition temperatures 130°C and 300°C for 60 min to give a 0.6 μm thick BST film. Sputtering pressure was varied between 5 and 12.5 mtorr in increments of 2.5 mtorr. Post deposition annealing was done in air at 650°C, 750°C and 900°C respectively to crystallize and densify the BST films. Annealing time was varied between 1 and 20 hours. The structural characterization of the films was done by a 4 circle Bruker AXS D-5000 diffractometer using CuK$_\alpha$ radiation source.

Standard photolithography and metal lift off process was used to define the structure of the interdigitated capacitors (IDCs) on the BST/ alumina sample. Lift off process was used since it utilizes benign chemicals which do not harm the BST thin films. This is in contrast to an etchback process where potentially harmful acids are used. Positive imaging photoresist Shipley 1813 and Microchem LOR5A was used to make a thick bilayer photoresist layer. After standard UV exposure and development of the photoresist the sample was ready for metallization. A thin layer of Cr (0.03 μm) was deposited first by magnetron sputtering at room temperature. Then 0.5 μm of Cu was deposited by thermal evaporation. Cr acts as an adhesion layer and Cu is the top electrode. Lift off was done in Microchem Remover PG solution to define the array of interdigitated electrodes. The IDCs had 6 fingers that are 3μm wide, 3μm apart, and 50 μm long.

A Hewlett Packard 4192 A LF impedance analyzer was used to measure the voltage dependent capacitance and loss tangent values of the BST IDCs .A voltage sweep was done from –35 V to +35 V in 2 V increments. The ac oscillation level for the electrical measurements presented here was 0.05 V. For all of these measurements a 1 MHz frequency was required since capacitance value only ~1 pF.

An Agilent (model E 4991A) RF impedance analyzer was used to study the frequency dependence of the capacitance and loss tangent of the IDCs. Cascade Microtech GS probes with a pitch of 150 μm was used for this purpose. The measurement setup was calibrated using a commercial standard. All measurements were done at room temperature between 1 MHz and 1 GHz.

A Keithley 617 programmable electrometer was used to measure the room temperature current voltage (I-V) characteristics. A voltage step of 1 V, a pre test relaxation time of 5s, and a delay time of 3s were used in all cases.

A Hewlett Packard 8510 C Network Analyzer was used to characterize the microwave properties (1 to 26.5 GHz) of the BST IDCs in a one-port configuration. Prior to testing, 100 μm pitch GS probes were calibrated using on - wafer standards. Reflection (S_{11}) data was treated using a model[13] which takes into account series resistance, inductance and a parallel resistor- capacitor (RC) circuit. Thus the device quality factor was determined which includes contributions from the dielectric, substrate, and metallization lines at GHz frequencies.

RESULTS AND DISCUSSION

The aim of this work is to demonstrate the ability to prepare high quality BST thin devices using both inexpensive materials and processes. A series of experiments were performed to identify the optimal preparation conditions such as sputtering conditions and post deposition anneals for BST thin films on alumina.

Fig. 1 shows a x-ray diffraction pattern of a BST film deposited on alumina at 300°C and then annealed in air at 900°C for 20 hours. It shows fully crystalline BST perovskite structure on alumina substrate.

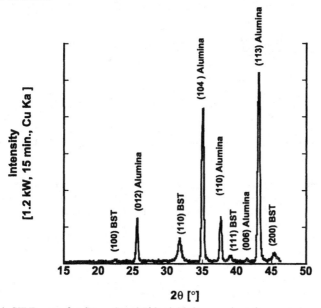

Figure 1. XRD scan for $(Ba_{0.75}Sr_{0.25})TiO_3$ / Al_2O_3 samples after post deposition anneal in air at 900°C for 20 hours. BST was deposited at 300°C under 10 mtorr sputtering pressure.

Figs. 2A through 2C show the dependency of tuning on sputtering pressure, deposition temperature, and post annealing temperature and time. As expected, higher temperatures and times improved tunability. For a 60 minute deposition, which corresponds to a 600 nm film, the optimum conditions were a 300°C substrate temperature, a 10 mTorr deposition pressure, and a 20 hour post deposition anneal at 900°C in air.

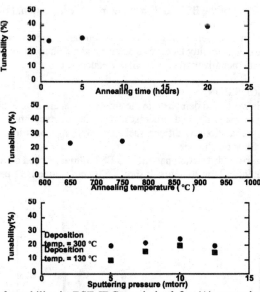

Figure 2.Variation of tunability in BST IDCs optimized for (A) sputtering pressure and BST deposition temperature with 650°C post deposition anneal, (B) post deposition annnealing temperature at 10 mTorr $P_{deposition}$, and 300°C T_{dep}, and (C) Post deposition anneal time at 10 mTorr deposition pressure 300°C T_{dep}, and 900°C T_{anneal}. The maximum applied field was 85 KV/cm.

Higher temperature post deposition anneals were not investigated because of film cracking upon cooldown, due to thermal expansion mismatch between BST and alumina. Also increasing the deposition temperature to 450°C did not improve tunability. These optimal conditions provided 40% tuning at 35 volts - the limit of our measurement instrumentation. Figs. 3 through 4 show voltage and frequency dependent responses for BST IDCs prepared using these optimal conditions. Fig. 3 shows the dielectric characteristics (capacitance and loss tangent vs. voltage) of the interdigitated capacitors. The capacitance changed from 0.375 pF at 0 V to 0.224 pF at 35 V. Percentage tunability is defined as $\{100 * (C_{(0 V)} - C_{(35 V)}) / C_{(0 V)}\}$ where $C_{(0 V)}$ and $C_{(35 V)}$ are the capacitance values at 0 V and 35 V respectively. Hence the tunability is 40 % at 35 V. The loss tangent (tanδ) is found to be 0.011 and hence a Q (quality factor) of 91 is obtained at O V. This loss tangent value decreases to 0.004 (Q = 250) for a bias of 35 V at 1 MHz.

Fig.4 shows the response of the BST IDCs as a function of frequency. We find that there is negligible change in the capacitance value (and hence the permittivity) as the frequency is increased from 1 MHz to 1 GHz. Thus the permittivity of the BST thin films can be considered frequency independent upto 1 GHz.

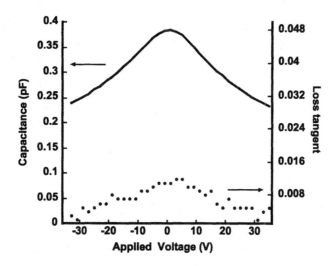

Figure 3.Capacitance and loss tangent as a function of applied voltage of the Cu / Cr / BST / Alumina interdigitated capacitors. All measurements performed at 1 MHz.

Fig.4. shows the loss tangent versus frequency plot. At 500 MHz the loss tangent value is 0.011which is the same as what we observe at lower frequency for zero bias. However close to 1 GHz the loss tangent value rises rapidly. This dispersion is mainly due to conductor losses, which begin to dominate at these frequencies and hence have a significant impact on the device quality factor. Hence the loss tangent of the BST films are also frequency independent[14].

An important criterion for assessing MW devices is quality factor measured over a higher range of frequencies, especially in GHz range where metallization resistance becomes increasingly important[15]. Thus the resistance of the electrodes has a significant contribution in the device loss at high frequencies. Hence the conductivity of the metal electrode and also its thickness plays an important role in determining the Q value at microwave frequencies. Microwave measurements were performed from 3 to 26 GHz and the device quality factor (including electrode resistance) was found to be ~30 at 26 GHz.

As shown in Fig.5, for an applied field of 100 kV/cm the leakage current is 1.39×10^{-10} A. This corresponds to a leakage current density of 1.0×10^{-6} A / cm^2 for an electric field of 100 kV/ cm of the BST planar varactors. Such low values of leakage current are necessary to achieve high reliability of BST based devices and will ultimately have a positive impact on the device lifetime.

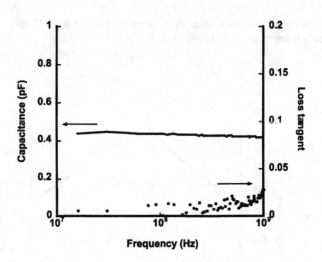

Figure 4.BST IDC loss tangent and capacitance vs. frequency at zero bias.

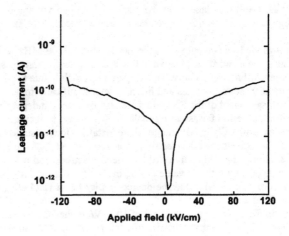

Figure 5. Leakage current vs. applied field for BST IDCs.

Finally, we compare our present data with several recent literature examples in Table 1. These data illustrate that this alumina-based technology is comparable, and in some cases superior to conventional substrate technologies.

	This work	Liu et al.[1]	Kim et al.[11]	Moon et al.[16]	Bellotti et al.[?]
Tunability (%)	40	26	64	40	65
Electric Field (V/μm)	11.6	40	35	13.3	7
Q (@ n GHz)	30 (26)	>20 (24)	27 (2.4)	9 (9)	4 (20)
	C = 0.6 pF	C = 7 pF	C = 3.5 pF	C = 0.4 pF	C = 1.9 pF

Table 1: Literature values for tunability, tuning field, and Q.values. Here the applied electric field is estimated by dividing the applied voltage by the IDC finger spacing.

SUMMARY

Cu/BST/Al$_2$O$_3$ thin film interdigitated capacitors were fabricated and evaluated between radio and microwave frequencies. Tunability of 40 % at 35 volts and a dielectric quality factor of ~ 100 are obtained at 1 MHz for zero bias. Device quality factors of ~30 are obtained at 26 GHz at zero bias. The BST thin films show low leakage current characteristics (I= 1.39 x 10^{-10} A) for an electric field of 100 kV/ cm. The dielectric constant and loss tangent of BST thin films are found to be frequency independent, while conductor losses begin to dominate device Q in the GHz range. We have demonstrated that tunable microwave devices based on BST thin films on low cost polycrystalline alumina substrates using single step Cu metallization looks very promising and compares well to devices fabricated using expensive single crystalline substrates and noble metallization.

ACKNOWLEDGEMENTS

This work was partially supported by a grant from NSF/ITR under contract no. 0113350 and work supported by US Army Communications and Electronics Command as a DARPA Grant through Purdue University under grant number DAAB07-02-1-L430.

REFERENCES
[1]Y. Liu, A. S. Nagra, E. G. Erker, P. Periaswamy, T. R. Taylor, J. Speck, and R. A. York, "BaSrTiO$_3$ Interdigitated Capacitors for Distributed Phase Shifter Applications," *IEEE Microwave And Guided Wave Lett.*, **10**, 448- 450 (2000)
[2]A. Tombak, J. P. Maria, F. T. Ayguavives, Z. Jin, G. T. Stauf, A. I Kingon, and A. Mortazawi, "Voltage – Controlled RF Filters Employing Thin – Film Barium - Strontium - Titanate Tunable Capacitors," *IEEE Trans.Microwave Theory Tech*, **51**, 462-467 (2003)
[3]Jayesh Nath, Dipankar Ghosh, Jon-Paul Maria, Michael B. Steer, and Angus I. Kingon," A Tunable Combline Bandpass Filter Using Thin Film Barium Strontium Titanate (BST)" - *Proceedings of the Asia Pacific Microwave Conference* (APMC), 939- 940, New Delhi, India (2004)

[4]C.B. Parker, J.P. Maria, and A.I. Kingon, "Temperature and thickness dependent permittivity of (Ba, Sr)TiO$_3$ thin films," *Appl. Phys. Lett.*, **81**, 340-342 (2002)

[5]D.M.Bubb, J.S.Horwitz, S.B.Qadri, S.W.Kirchoeffer, C.Hubert, J.Levy, "(Ba, Sr)TiO$_3$ thin films grown by pulsed laser deposition with low dielectric loss at microwave frequencies,"*Appl. Phys. A*, **79**, 99 - 101(2004)

[6]J.C.Jiang, Y.Lin, C.L.Chen, C.W.Chu, E.I.Meletis, "Microstructure and surface step-induced antiphase boundaries in epitaxial ferroelectric Ba$_{0.6}$Sr$_{0.4}$TiO$_3$ thin film on MgO," *J. Appl. Phys.*, **91**, 3188 – 3192 (2002)

[7]A.Srivastava, D.Kumar, R.K.Singh, H.Venkataraman, and W.R.Eisenstadt, "Improvement in electrical and dielectric behavior of (Ba, Sr) TiO$_3$ thin films by Ag doping," *Phys. Rev. B*, **61**, 7305- 7307(2000)

[8]S.I. Jang and H.M. Jang, "Structure and electrical properties of boron – added (Ba, Sr)TiO$_3$ thin films fabricated by the sol – gel method," *Thin Solid Films*, **330**, 89 - 95 (1998)

[9]J. Bellotti, E. K.Akdogan, A. Safari, W. Chang, S. Kirchoefer, "Tunable Dielectric Properties of BST Thin Films for RF/MW Passive Components,"*Integr. Ferroelectr.*, **49**, 113 - 122 (2002)

[10]J.Xu, W. Menesklou, E.I. Tiffee, "Processing and properties of BST thin films for tunable microwave devices," *J. Eur. Cer. Soc.*, **24**, 1735 - 1739 (2004)

[11]D. Kim, Y. Choi, M. Ahn, M.G. Allen, J.S. Kenney, and P. Marry, "2.4 GHz Continuously Variable Ferroelectric Phase Shifters Using All – Pass Networks," *IEEE Microwave Wireless Components Lett.*, **13**, 434 -436 (2003)

[12]W.Fan, B.Kabius, J.M.Hiller, S.Saha, J.A.Carlisle, O.Aucielle, R.P.H. Chang, R.Ramesh, "Materials science and integration bases for fabrication of (Ba$_x$Sr$_{1-x}$TiO$_3$) thin film capacitors with layered Cu - based electrodes," *J. Appl. Phys.*, **94**, 6192- 6200 (2003)

[13]J.Nath, D.Ghosh, J.P.Maria, M.B.Steer, A.I.Kingon, and G.T.Stauf, "Microwave properties of BST thin film interdigital capacitors on low cost Alumina substrates," *Proceedings of the 34th European Microwave Conference*, 1497-1500, Amsterdam, Netherlands, 11-15 Oct (2004)

[14] Z.Jin, A.Tombak, J. – P. Maria, B.Boyette, G.T.Stauf, A.I.Kingon, and A.Mortazawi, "Microwave Characterization of Thin Film BST Material Using a Simple Measurement Technique," *IEEE MTT – S International Microwave Symposium Digest*, **2**, 1201-1204 (2002)

[15]D.C. Dube, J. Baborowski, P.Muralt, and N.Setter, "The effect of bottom electrode on the performance of thin film based capacitors in the gigahertz region,"*Appl. Phys. Lett.*, **74**, 3546 – 3548, (1999)

[16]S.E. Moon, E- Y. Kim, M- H. Kwak, H – C. Ryu, Y- T. Kim, K – Y. Kang, S – J. Lee, and W – J. Kim, "Orientation dependent microwave dielectric properties of ferroelectric Ba$_{1-x}$Sr$_x$TiO$_3$ thin films," *Appl. Phys. Lett.*, **83**, 2166-2168(2003)

Microwave Dielectric Materials

IONIC DISTRIBUTION AND MICROWAVE DIELECTRIC PROPERTIES FOR TUNGSTENBRONZE-TYPE LIKE Ba$_{6-3x}$R$_{8+2x}$Ti$_{18}$O$_{54}$ (R = Sm, Nd and La) SOLID SOLUTIONS

H. Ohsato, M. Suzuki and K. Kakimoto
Materials Science and Engineering, Nagoya Institute of Technology, Gokiso-cho, Showa-ku, Nagoya 466-8555, Japan

Ba$_{6-3x}$R$_{8+2x}$Ti$_{18}$O$_{54}$ (R = Sm, Nd and La) solid solutions have been applied to microwave resonators and filters for mobile phone because of their high dielectric constant ε_r, high quality factor $Q \cdot f$ and near zero temperature coefficient of resonant frequency τ_f. In this paper, relationships between cation distribution and microwave dielectric properties in this compound are presented in four following cases: case 1 is the improvement of quality factor by compositional ordering; case 2 does more compositional ordering result in higher quality factors; case 3 the quality factor dependence on R-cations; and case 4 the correlation of trigonal site occupation with microwave dielectric properties.

INTRODUCTION

Nowadays, microwave telecommunication technology has been developed for the following applications: portable telephone, satellite broadcasting, ultra-high speed wireless LAN, intelligent transport system (ITS) including ETC and ladder for anticollision, and so on. Further development of microwave dielectrics is necessary for wireless communications. The tungstenbronze-type like Ba$_{6-3x}$R$_{8+2x}$Ti$_{18}$O$_{54}$ solid solutions located on the tie-line between BaTiO$_3$-R$_2$Ti$_3$O$_9$ composition as shown in Fig.1 have been applied to resonators and filters in mobile phones because of their high dielectric constant.[1,2] This compound has been also used for low temperature cofired ceramics (LTCC)[3] as shown in Table I .

The Ba$_{6-3x}$R$_{8+2x}$Ti$_{18}$O$_{54}$ solid solutions have a tungstenbronze-type like structure as shown in Fig. 2. The crystal data with superlattice doubling of the c-axis are as follows: orthorhombic Pbnm (No.62), a=12.131(13), b=22.271(5), c=7.639(5) Å and Z=2[4]. The fundamental unit cell of the structure[1] contains three types of large cation sites: ten A1-rhombic sites in 2 × 2 perovskite blocks, four A2-pentagonal sites, and four C-trigonal sites. The pentagonal and trigonal sites are located among the perovskite blocks. The fundamental structure is expressed by the formula

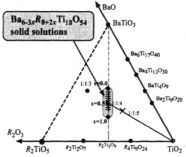

Fig. 1 Ba$_{6-3x}$R$_{8+2x}$Ti$_{18}$O$_{54}$ solid solutions in BaO-R$_2$O$_3$-TiO$_2$ ternary system.

Table I Microwave dielectric properties of tungstenbronze $Ba_{6-3x}R_{8+2x}Ti_{18}O_{54}$ solid solutions.

	ε_r	$Q\cdot f$ GHz	τ_f ppm/°C
For high $Q\cdot f$			
Sm- system	81	10543	-11.3
Nd-system	88	10010	76
For near zero τ_f			
Sm,Nd-system	85	9160	8.6
Sm, La-system	84	9046	1.6
For LTCC[3]			
$BaO-Nd_2O_3-Bi_2O_3-TiO_2$	81	>3000	-10
$BaO-Nd_2O_3-TiO_2$	85	>3600	-

$[R_{8+2x}Ba_{2-3x}V_x]_{A1}$ $[Ba_4]_{A2}[V_4]_CTi_{18}O_{54}$ ($0\leq x \leq 2/3$)[5, 6]. Here, V means vacancy. In the $0 \leq x \leq 2/3$ composition region, the $A1$-sites are occupied mainly by medium-sized Sm-ions, and also by a small amount of large Ba-ions. For $x = 2/3$, the $A1$-sites and $A2$-sites are occupied by R-ions and Ba-ions, respectively. On the other hand, the C-sites are unoccupied by cations, because they are the smallest sites.

The coordination numbers (CN) and configurations of each polyhedron in the tungstenbronze-type-like $Ba_{6-3x}R_{8+2x}Ti_{18}O_{54}$ crystal structure are illustrated in Fig. 3. $A1(1)$ and $A1(5)$-sites are two-cap trigonal prisms with 8 CN, $A1(2)$- and $A1(4)$-sites are distorted cubic dodecahedra with 8 CN, and the $A1(3)$-site is a three-cap trigonal prism with 9 CN, whereas $A2(1)$- and $A2(2)$-sites are two-cap hexahedra with 10 CN. The configurations of C-sites have not been determined yet because of non crystal structure analysis containing some ions in C-sites.

In this study, relationships between cation distribution and microwave dielectric properties have been studied for the following four cases. For case 1 the effect of compositional ordering on the quality factor is presented, while for case 2 the effects of enhanced compositional ordering is considered. Case 3 is the effect of R cation species on the quality factor and case 4 considers the effect of trigonal site occupation on the quality factor.

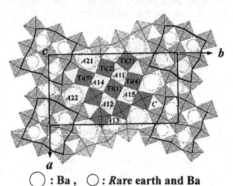

$\bigcirc$: Ba , $\bigcirc$: Rare earth and Ba

Fig. 2 Crystal structure of tungstenbronze $Ba_{6-3x}R_{8+2x}Ti_{18}O_{54}$ solid solutions.

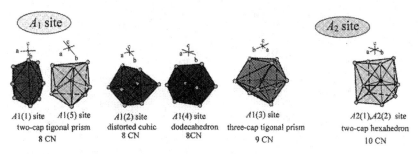

$A1(1)$ site $A1(5)$ site $A1(2)$ site $A1(4)$ site $A1(3)$ site $A2(1),A2(2)$ site
two-cap tigonal prism   distorted cubic   dodecahedron   three-cap tigonal prism   two-cap hexahedron
8 CN   8 CN   8CN   9 CN   10 CN

Fig. 3 Configuration of polyhedra of cation sites.

EXPERIMENTAL PROCEDURE

Crystal Structure

Crystal data were obtained with aid of single crystal X-ray diffraction using the Weissenberg camera and precession camera. The single crystals were grown by self-flux method. The superlattice was found using an oscillation photograph as reported in previous paper[4].

Diffraction intensity data were collected by using a single crystal four-circle X-ray diffractometer and Weissenberg-type X-ray diffractometer with imaging plate with graphite-monochromatized MoKα radiation. The initial atomic parameters for the fundamental lattice were used from Matveeva et al.[5] and those of superlattice were derived from those of the fundamental lattice. The intensity data were corrected for Lorentz and polarization factors and absorption. Full-matrix least-squares refinement on F of atomic parameters, with anisotropic thermal parameters and occupations of rare earth ions, was carried out with program $RADY$[6].

Properties of Microwave Dielectrics

The samples for the measurement of the microwave dielectric properties, the internal strain and lattice parameters were prepared by a solid state reaction as reported in previous papers[7]. The dielectric properties were measured by the Hakki and Coleman method[8, 9]. The samples with optimal $Q \cdot f$ values were selected for discussion.

The internal strain η was obtained from the following equation[10] as the grain size of the ceramics is sufficiently large:

$$\beta = 2\eta \tan\theta.$$

Here, β is the full-width at half-maximum (FWHM) of the X-ray powder diffraction peaks. The powder patterns were obtained by multi-detector system (MDS)[11] using synchrotron radiation in the "Photon Factory" of the National Laboratory for High Energy Physics in Tsukuba, Japan. The more precise measuring conditions were reported in previous paper.[7]

Accurate lattice parameters were obtained using the whole-powder-pattern-decomposition method (WPPD) program[12] for the powder diffraction patterns obtained by the step scanning technique using Si (99.99%) as an internal standard.

RESULTS AND DISCUSSION

Case 1: Quality Factor Improvement by Compositional Ordering[7]

137

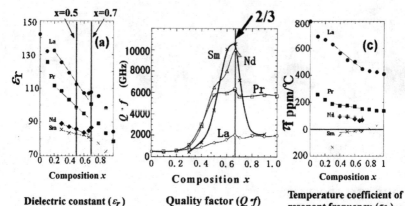

Dielectric constant (ε_r) Quality factor ($Q \cdot f$) Temperature coefficient of resonant frequency (τ_f)

Fig.4 Microwave dielectric properties of tungstenbronze-type like $Ba_{6-3x}R_{8+2x}Ti_{18}O_{54}$ solid solutions.

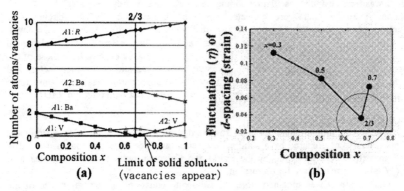

(a) Limit of solid solutions (vacancies appear) (b)

Fig.5 Compositional ordering. (a) Number of atoms and vacancies in large cations site. (b) Internal strain/ fluctuation of d-spacing as a function of composition x.

The dielectric properties as a function of the composition x for the $Ba_{6-3x}R_{8+2x}Ti_{18}O_{54}$ solid solutions[13, 14] are shown in Fig. 4. Similar results for the Sm and Nd system are shown by Negas et al.[15]. The characteristic phenomenon is that $Q \cdot f$ values varied non linearly as a function of composition, though ε_r and τ_f vary proportional to the composition. The data of the $x=2/3$ composition show the highest $Q \cdot f$ values: 10549GHz in the Sm system, 10010GHz in the Nd system and 2024GHz in the La system as shown in Fig.4.[7] Two Ba ions occupy $A1$-sites at $x=0$ as shown in the structural formula $[R_{8+2x}Ba_{2-3x}V_x]_{A1}[Ba_4]_{A2}Ti_{18}O_{54}$ in the range of $0 \leq x \leq 2/3$. The amount of Ba ions in $A1$-site is decreased as increasing x, and at $x=2/3$ the number of Ba ions becomes zero. The highest quality factor is based on the ordering of R- and Ba-ions in the $A1$- and $A2$-sites, respectively.

The ordering of the cations might reduce the internal strain η / fluctuation of lattice d-spacing[16]. In Fig.5, $Q \cdot f$ are shown as a function of η. Up to $x=2/3$, $Q \cdot f$ increases linearly with decreasing η. The internal strain at just $x=2/3$ is the lowest. This low internal strain comes from the distribution of cations in the rhombic sites and the pentagonal sites.

In the $x=2/3$ composition, ions with the same size occupy each $A1$ and $A2$ site, as shown in the structural formula $[R_{9.33}V_{0.67}]_{A1}[Ba_4]_{A2}Ti_{18}O_{54}$, that means, R-ions and Ba-ions are ordering in both the rhombic sites and pentagonal sites, respectively. This ordering leads to the lowest strain. On the other hand, as the x-values increase according to the structural formula $[R_{9.33+2(x-2/3)}V_{0.66-(x-2/3)}]_{A1}$ $[Ba_{4-3(x-2/3)}V_{3(x-2/3)}]_{A2}Ti_{18}O_{54}$ in the range of $2/3 \leq x \leq 0.7$, Ba ions in $A2$-sites are substituted by R-ions. The decrease of Ba-ions produces vacancies in $A2$-sites and may lead to unstable crystal structures, as shown by the limit of solid solubility located near the $x=0.7$ composition.[17]

Case 2: More Compositional Ordering [18, 19]

Sr-ions are introduced in this system, for which the ionic size is between that of Ba- and Sm-ions. As mentioned above, $Q \cdot f$ values of those $Ba_{6-3x}R_{8+2x}Ti_{18}O_{56}$ solid solutions have the maximum value at $x=2/3$. In the region of smaller than $x = 2/3$, the structural formula of the solid solutions is $[R_{8+2x}Ba_{2-3x}V_x]_{A1}[Ba_4]_{A2}[V_4]_C Ti_{18}O_{54}$. In this region, Ba-ions locate in $A1$-sites, which deteriorate the quality factor. Following two cases are presented in this section:

At $x=0$, $[R_8 Ba_2]_{A1}[Ba_4]_{A2}[V_4]_C Ti_{18}O_{54}$ in which all $A1$- and $A2$-sites are occupied by Ba- and R-ions.[18]

At $x=0.6$, $[R_{9.2}Ba_{0.2}V_{0.6}]_{A1}[Ba_4]_{A2}[V_4]_C Ti_{18}O_{54}$ [19]

In the case of $x=0$, $Q \cdot f$ values are very low as shown in Fig. 4. When Ba-ions are substituted by Sr-ions such as $[R_8 Sr_2]_{A1}[Ba_4]_{A2}[V_4]_C Ti_{18}O_{54}$, $Q \cdot f$ values are improved steeply from 206 to 5880GHz as shown in Fig. 6. Introduction of Sr-ions in $A1$-sites may reduce the internal stress/ fluctuation of d-spacing, because of reducing ionic size in $A1$-sites. Mercurio et al.[20] reported that Sr-ions occupy $A13$ special sites which are medium size between $A1$- and $A2$-sites. So, it is thought that R, Sr and Ba-ions are ordering in $A1$-, $A13$- and $A2$-sites, respectively.

In the case of $x=0$: Qf value is very low   **Qf value is increased steeply**

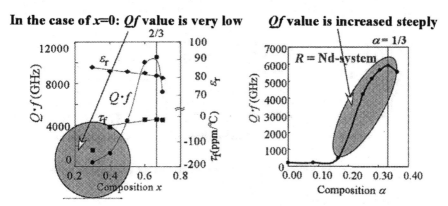

Fig.6 More compositional ordering. (a) $Q \cdot f$ as a function of x. (b) $Q \cdot f$ as a function of α in $(Ba_{1-\alpha}Sr_\alpha)_6 Nd_8 Ti_{18}O_{54}$.

In the case of x=0.6, the structural formula is $[R_{9.2}Ba_{0.2}V_{0.6}]_{A1}[Ba_4]_{A2}[V_4]_cTi_{18}O_{54}$. Number of Ba-ions in A1-sites is 0.2 atoms. When these Ba-ions are substituted by Sr-ions, the structural formula is $[R_{9.2}Ba_{0.2-\alpha}Sr_\alpha V_{0.6}]_{A1}[Ba_4]_{A2}[V_4]_cTi_{18}O_{54}$ in the range of $0.0 \leq \alpha \leq 0.2$ and $[R_{9.2}Sr_{0.2+(\alpha-0.2)}V_{0.6-(\alpha-0.2)}]_{A1}[Ba_{4-(\alpha-0.2)}]_{A2}[V_4]_cTi_{18}O_{54}$ in the range of $0.2 \leq \alpha < 0.8$. The lattice parameters as shown in Fig. 7 reveal the structural formula is reasonable. Up to α=0.2 the lattice parameter decreases according to the size differences between Ba- and Sr-ion. After that, the lattice parameters increase because vacancies in A1-sites are occupied by Sr-ions substituted for Ba-ions in A2-sites. The microwave dielectric properties also are affected by the occupation of cations as shown in Fig. 8. $Q \cdot f$ products increase from 9500GHz to 10200GHz up to α=0.2. This increase depends on the substitution of Sr- for Ba-ions with decreasing inner strain, which is caused by the larger Ba ions occupying the A1-sites. In the range of more than α=0.2, the $Q \cdot f$ decreases gradually to 3000GHz, because of the vacancies generated by the substitution the Sr- for Ba-ions in A2-sites. The vacancy brings the crystal structural unstable. The dielectric constants have also a turning point in the composition at α=0.2. Here, ε_r is minimum, because reducing of the size of the TiO_2 oxygen octahedral depends on the decrease in the lattice parameter. The temperature coefficient of resonant frequency τ_f behaves like the change in the ε_r as reported for many dielectric materials.[21, 22] However, the reason for this has not yet been clarified.

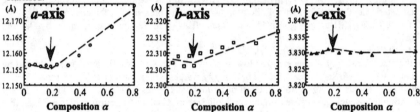

Fig.7 Lattice parameters substituted for Ba by Sr-ions according to the structural formula $[R_{9.2}Ba_{0.2-\alpha}Sr_\alpha V_{0.6}]_{A1}[Ba_4]_{A2}[V_4]_cTi_{18}O_{54}$ in the range of $0.0 \leq \alpha \leq 0.2$ and $[R_{9.2}Sr_{0.2+(\alpha-0.2)}V_{0.6-(\alpha-0.2)}]_{A1}[Ba_{4-(\alpha-0.2)}]_{A2}[V_4]_cTi_{18}O_{54}$ in the range of $0.2 \leq \alpha < 0.8$.

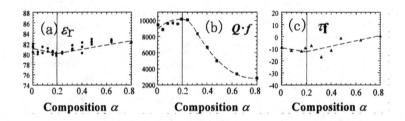

Fig.8 Microwave dielectric properties obtained by substitution of Ba by Sr-ions.

Case 3: Quality Factor Depending on R-Cations[23, 24]

The dielectric properties of each R-compound with x=2/3 are shown as a function of ionic radius as shown in Fig. 9. It should be noticed that the quality factors ($Q \cdot f$) increase with the decrease of the ionic radii of each R-ion as described later. The $Q \cdot f$ values of Sm- and Nd-compounds are excellent. The dielectric constants ε_r decrease linearly from 105 to 80 as a function of ionic radius, which are inversely proportional to the $Q \cdot f$ values as generally observed in dielectric properties. The ε_r of La-compound is the highest due to a decrease of tilt angle of TiO_6-octahedron as follows from the increase of lattice parameter. The temperature coefficients of the resonant frequency τ_f decrease proportionally to ε_r. Note that the τ_f of the Sm-compound is close to 0 ppm/°C. Totally, the dielectric properties of Sm- and Nd-compounds are most appropriate for application in the microwave resonator.

The $Q \cdot f$ of the Sm-compound is the highest due to the low internal strain/ fluctuation of d-spacing on the atomic scale as shown in Fig. 9. It should be noticed that h for Sm is the lowest. This low h originates from the difference of ionic radii between R- and Ba-ions. In the case of $Ba_{6-3x}R_{8+2x}Ti_{18}O_{54}$ solid solution series, lower internal strain/ fluctuation of d-spacing originates from the ordering of R- and Ba-ions that occupy $A1$- and $A2$-sites, respectively. In the case of R-compounds with $x=2/3$, we are comparing R-compounds with different sizes of ionic radii. The tungstenbronze-type like structure is built with two different parts. One is perovskite blocks which include medium size ions such as R-ions, and another is pentagonal columns, which include large size ions such as Ba-ion. These two parts are produced by existence of different size in cations. Therefore, the size of cations occupying the two sites should be different. Sm-compound with the smallest ionic radius in the R-compound series is the most stable for the rhombic $A1$-sites in the perovskite blocks, which show the smallest internal strain. This stabilization of the crystal structure has improved $Q \cdot f$ values.

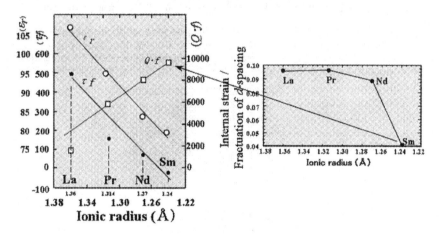

Fig,9 Quality factor and internal strain/ fractuation of d-spacing depend on the R-ions.

Case 4: Trigonal Site Occupation[25]

Li-ions with smaller ionic radius than the Sm-ion are introduced in this system for occupation to C-sites. A series of $Ba_4Sm_{(28-y)/3}Li_yTi_{18}O_{54}$ solid solutions with three Li-ions substituted for one Sm-ion is examined. The lattice parameters of the sintered samples as a function of the composition y are shown in Fig. 10. The gradient of the lattice parameters changes at $y = 1$. The lattice parameters of the a- and b-axes decrease steeply in the first region $0 \leq y \leq 1$, and increase linearly in the next region $1 \leq y \leq 7$. For $y > 7$, the lattice parameters are unchanged. In the first two stages, the lattice parameters change linearly according to the Vegard's rule. The two stages form different types sub-solid solution series in the $Ba_4Sm_{(28-y)/3}Li_yTi_{18}O_{54}$ solid solutions $(0 \leq y \leq 7)$.

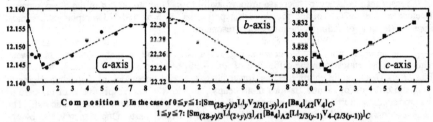

Composition y In the case of $0 \leq y \leq 1$: $[Sm_{(28-y)/3}Li_yV_{2/3(1-y)}]_{A1}[Ba_4]_{A2}[V_4]_C$

$1 \leq y \leq 7$: $[Sm_{(28-y)/3}Li_{(2+y)/3}]_{A1}[Ba_4]_{A2}[Li_{2/3(y-1)}V_{4-(2/3(y-1))}]_C$

Fig.10 Lattice parameters of tungstenbronze-type like $Ba_4Sm_{(28-y)/3}Li_yTi_{18}O_{54}$ solid solutions.

In the early stage $0 \leq y \leq 1$, the lattice parameters decreased, indicating that Li-ions occupied only $A1$-sites. The structural formula is $[Sm_{(28-y)/3}Li_yV_{2/3(1-y)}]_{A1}[Ba_4]_{A2}[V_4]_C Ti_{18}O_{54}$. In this region, one Sm-ion in the $A1$-sites is substituted by three Li-ions. In the composition $y = 1$, the structural formula is $[Sm_9Li_1]_{A1}[Ba_4]_{A2}[V_4]_C Ti_{18}O_{54}$. Here, Sm and Li-ions occupy all the ten $A1$-sites, sharing up nine sites and one site, respectively, so that no vacancies remain in the $A1$-sites.

In the subsequent stage $1 \leq y \leq 7$, the lattice parameters of the a- and b-axes increased, indicating that Li-ions occupied the C-sites. The formula is $[Sm_{(28-y)/3}Li_{(2+y)/3}]_{A1}[Ba_4]_{A2}[Li_{2/3(y-1)}V_{4-(2/3(y-1))}]_C Ti_{18}O_{54}$. Li-ions start to occupy the trigonal C-sites after filling all of the $A1$-sites, because the substitution of three Li-ions for one Sm-ion continues beyond the

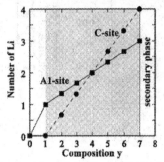

Fig.11 Number of Li-ions in $A1$- and C-sites of tungstenbronze-type like $Ba_4Sm_{(28-y)/3}Li_yTi_{18}O_{54}$ solid solutions.

composition $y = 1$. Finally, at $y = 7$, the structural formula is $[Sm_7Li_3]_{A1}[Ba_4]_{A2}[Li_4]_C Ti_{18}O_{54}$, in which all of the cation sites are occupied by Sm, Ba and Li-ions. This final composition is coincident with the limiting point of the solid solutions. The numbers of Li-ions in the $A1$- and C-sites are shown in Fig.11 as a function of y. The present substitution study was undertaken in order to understand which site the Li-ions with small ionic radii would first occupy. The C-sites which remained empty in the crystal structure of the $Ba_{6-3x}R_{8+2x}Ti_{18}O_{54}$ solid solutions are expected to be occupied by small ions. To the best of our knowledge, this crystallographic hypothesis could not be confirmed. We concluded that Li-ions occupied the $A1$-sites during the first region, before they occupied the C-sites, according to the lattice-parameter changes described before. The occupation tendencies of the Li-ions resulted from size differences among the $A1$-sites, the C-sites, and the Li-ions.

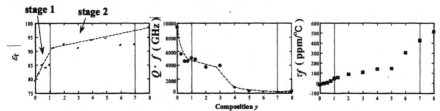

Fig.12 Microwave dielectric properties of $Ba_{6-3x}Sm_{8+2x}Ti_{18}O_{54}$ solid solutions substituted for Sm-ions by Li-ions.

The microwave dielectric properties—dielectric constant ε_r, quality factor $Q \cdot f$, and temperature coefficient of the resonant frequency τ_f are shown as a function of composition y in Fig. 12. These properties correlate with the lattice parameter changes shown in Fig. 10. Near the composition $y = 1$, the behavior of the microwave properties change as a function of y. In particular, ε_r increases linearly in the stage of $0 \leq y \leq 1$, despite the decreasing lattice parameters. Usually, the ε_r values are proportional to the lattice parameters, because the ionic polarizabilities depend on the size of the TiO_6 octahedron, as reported for the $Ba_{6-3x}R_{8+2x}Ti_{18}O_{54}$ solid solutions in a previous paper.[26] In this same stage, $0 \leq y \leq 1$, $Q \cdot f$ values decrease steeply with the substitution of Li for Sm-ions. The decrease in $Q \cdot f$ value results from the increased polarizabilities and the internal strain[7] caused by the substitution of lithium ions. The τ_f value also increases proportionally with ε_r, as observed in a previous work,[26] but the reason for that increase has not yet been clarified. τ_f becomes 0 ppm/°C in the vicinity of $y = 0.3$, accompanied by $\varepsilon_r = 83$ and $Q \cdot f$ = 5,000 GHz.

In the stage of $y > 1.0$, the ε_r and τ_f values increase linearly as a function of y, as shown by the dotted line in Fig. 12. In the whole region, the change in ε_r shows the same tendency with the number of Li-ions occupying the $A1$-sites as shown in Fig. 11. Thus, an increase in ε_r depends on the number of Li-ions occupying the $A1$-sites. However, as the radius of Li-ion is smaller than that of Sm-ion, the Li-ions contribute to ionic polarization, similar to that of the Ti-ions in the perovskite structure.

CONCLUSIONS

The relationships between cation distribution and microwave dielectric properties in tungstenbronze-type like $Ba_{6-3x}R_{8+2x}Ti_{18}O_{54}$ solid solutions are presented for four cases.

Case 1 is compositional ordering: Compositional ordering of rhombic $A1$- and pentagonal $A2$-sites improved the quality factor $Q \cdot f$. The maximum $Q \cdot f$ value was achieved by ordering of Ba and R-ions at $x = 2/3$.

Case 2 considers a higher degree of compositional ordering: more compositional ordering in $A1$-sites further improves the quality factor. When Ba-ions in $A1$-sites are substituted for by Sr-ions, the $Q \cdot f$ values are improved, because the compositional ordering of R, Sr, and Ba is formed in $A1$-, $A13$- and $A2$-sites, respectively.

Case 3 considers the quality factor dependence on the type of R-cations: R=Sm system had the highest ordering ratio of Ba- and R-ions, the second highest ordering is for the R=Nd system and the third highest ordering is for the R=La system. The difference of ionic radius between Ba- and R-ions increase the ordering ratio. Vacancies in $A2$-sites bring low quality factor because of unstable on the crystal structure.

Case 4 is the correlation of trigonal site occupation with microwave properties: in the early stage Li-ions occupy the $A1$-sites. In the subsequent stage, they occupy the trigonal C-site. Occupation of the trigonal C-site results in higher dielectric constants but lower $Q \cdot f$ products.

ACKNOWLEGHEMENTS

The authors would like to thank Mr. Masaki Imaeda, Mr. Atushi Komura, Yosuke Futamata. Mr. Mr. Hideaki Sakashita, Motoaki Imaeda, Mr. Hiromichi. Kato graduated in Nagoya Institute of Technology, and Mr.Tadashi Otagiri and Mr. Takashi Nagatomo in Soshin Electric co., Ltd., for collaborations on this study. The authors also would like to thank Dr. Susumu Nishigaki and Mr. Akio Harada, President of Daiken Chemical Co., Ltd., for assist of this study. A part of this study was supported by following projects: The NITECH 21st Century COE (Center of Excellence) program "World Ceramics Center for Environmental Harmony" supported by Japanese Ministry of Education, Science and Culture, and "NIT-Seto Ceramics R & D Project" supported by Japanese Ministry of Education, Science and Culture, and NEDO foundation for matching fund.

REFERENCES

[1] H. Ohsato, "Science of Tungstenbronze-type Like $Ba_{6-3x}R_{8+2x}Ti_{18}O_{54}$ (R = rare earth) Microwave Dielectric Solid Solutions," *J. Eur. Ceram. Soc.*, 21, 2703-2711 (2001).

[2] H. Ohsato, H. Kato, M. Mizuta, S. Nishigaki and T. Okuda, "Microwave Dielectric Properties of the Ba_{6-3x} $(Sm_{1-Y}, R_Y)_{8+2x}Ti_{18}O_{54}$ (R=Nd and La) Solid Solutions with Zero Temperature Coefficient of the Resonant Frequency", *Jpn. J. Appl. Phys.*, 34, 9B, 5413-5417 (1995).

[3] T. Otagiri, K. Kawamura and M. Inoue, "Development of the LTCC for Microwave Wireless Communication", *Function & Materials*, 24[6], 38-63 (2004).

[4] H. Ohsato, S. Nishigaki and T. Okuda, "Superlattice and Dielectric Properties of Dielectric Compounds", Jpn. J. Appl. Phys., 31, 9B, 3136-3138 (1992).

[5] R. G. Matveeva, M. B. Varforomeev, and L. S. ll'yuschenko, "Refinement of the Composition and Crystal Structure of $Ba_{3.75}Pr_{9.5}Ti_{18}O_{54}$," *Zh. Neorg. Khim.*, 29, 31-34 (1984), *Translation, Russ. J. Inorg. Chem.*, 29, 17-19 (1984).

[6] S. Sasaki, "A Fortran Program for the Least-Squares Refinement of Crystal Structures," *XL Report, ESS*, State Univ. of New York, 1-17 (1982)

[7] H. Ohsato, M. Imaeda, Y. Takagi, A. Komura, and T. Okuda, "Microwave Quality Factor Improved by Ordering of Ba and Rare-earth on the Tungstenbronze-type $Ba_{6-3x}R_{8+2x}Ti_{18}O_{54}$ (R=La, Nd and Sm) Solid Solutions," *Proceeding of the XIth IEEE International Symposium on Applications of Ferroelectrics*, IEEE catalog number 98CH36245, 509-512 (1998).

[8] B. W. Hakki, and P. D. Coleman, "A Dielectric Resonator Method of Measuring Inductive in the Millimeter Range," *IRE Trans. Microwave Theory & Tech.*, MTT-8, 402-410 (1960).

[9] Y. Kobayashi and M. Katoh, "Microwave Measurement of Dielectric Properties of Low-loss Materials by the Dielectric Resonator Method," *IEEE Trans. on MTT*, MTT-33, 586-92 (1985).

[10] A. R. Stokes, and A. J. C. Willson, "The Diffraction of X rays by Distorted Crystal Aggregates-I," *Proc. Phys. Soc.*, 56, 174-181 (1944).

[11] H. Toraya, H. Hibino, and K. Ohsumi, "A New Powder Diffractometer for Synchrotron

Radiation with Multiple-Detector System," *J. Synchrotron Rad.*, **3**, 75-83 (1996).

[12] H. Toraya, "Whole-Powder-Pattern Fitting without Reference to a Structural Model: Application to X-ray Powder Diffractometer Data," *J. Appl. Cryst.*, **19**, 440-447 (1986).

[13] H. Ohsato, T. Ohhashi, H. Kato, S. Nishigaki, and T. Okuda, "Microwave Dielectric Properties and Structure of the $Ba_{6-3x}Sm_{8+2x}Ti_{18}O_{54}$ Solid Solutions," *Jpn. J. Appl. Phys.*, **34**, 187-191 (1995).

[14] H. Ohsato, M. Mizuta, T. Ikoma, Z. Onogi, S. Nishigaki, and T. Okuda, "Microwave Dielectric Properties of Tungstenbronze-type $Ba_{6-3x}R_{8+2x}Ti_{18}O_{54}$ (R = La, Pr, Nd and Sm) Solid Solutions," *J. Ceram. Soci. Japan*, **106(2)**, 178-182 (1998).

[15] T. Negas, and P. K. Davies, "Influence of Chemistry and Processing on the Electrical Properties of $Ba_{6-3x}Ln_{8+2x}Ti_{18}O_{54}$ Solid Solutions," *Material and Processes for Wireless Communications. Ceramic Transactions.*, **53**, 197-196. (1995).

[16] H. Ohsato, A. Kan, K. Kakimoto, H. Ogawa and S. Nishigaki, "Microwave Dielectric Properties Correlated to Crystal Structure: An Analysis", *Proceedings of the 2002 13th IEEE Internationals Symposium on Applications of Ferroeletrics (ISAF 2002)*, Editors: Gray White and Takaaki Tsurumi, 72-78, Nara, Japan, May 28- June 1, 2002.

[17] H. Ohsato, T. Ohhashi, S. Nishigaki, T. Okuda, K. Sumiya and S. Suzuki, "Formation of Solid Solutions of New Tungsten Bronze-Type Microwave Dielectric Compounds $Ba_{6-3x}R_{8+2x}Ti_{18}O_{54}$ (R=Nd and Sm, 0<x<1)", *Jpn. J. Appl. Phys.*, **32, 9B**, 4324-4326 (1993).

[18] T. Nagatomo, T. Otagiri, M. Suzuki and H. Ohsato, "Microwave Eelectric Properties and Crystal Structure of the Tungstenbronze-Type Like $(Ba_{1-\alpha}Sr_{\alpha})6(Nd_{1-\beta}Y\beta)_8Ti_{18}O_{54}$ Solid Solutions", *J. Eur. Ceram. Soc.*, **25**, (2005) Accepted.

[19] M. Imaeda, M. Mizuta, K. Ito, H. Ohsato, S. Nishigaki and T. Okuda, "Microwave Dielectric Properties of Ba6-3xR8+2xTi18O54 Solid Solutions Substituted Sr for Ba", *Jpn. J. Appl. Phys.*, **36, 9B**, 6012-6015 (1997).

[20] D. Mercurio, M. Abou-Salama and J.-P. Mercurio, "Investigations of the Tungsten-bronze-Type $(Ba_{1-\alpha}Sr_{\alpha})_6La_{8+2x/3}Ti_{18}O_{54}$ ($0 \leq x \leq 3$) Solid Solutions" *J. Eur. Ceram. Soc.*, **21**, 2713-2716 (2001).

[21] A. J. Bosman and E. E. Havinga, *Phys. Rev.*, **129**, 1593 (1963).

[22] Harrop, *J. Mater. Sci.*, **4**, 370 (1969).

[23] H. Ohsato, M. Mizuta and T. Okuda, "Crystal Structure and Microwave Dielectric Properties of Tungstenbronze-type $Ba_{6-3x}R_{8+2x}Ti_{18}O_{54}$ (R = La, Pr, Nd and Sm) Solid Solutions," *Applied Crystallography*, 440-447 (1997).

[24] A. Kan, H. Ogawa and H. Ohsato, "Microwave Dielectric Properties of $R_2Ba(Cu_{1-x}M_x)O_5$(R=Y and Yb, M=Zn and Ni) Solid Solutions", *Materials Chemistry and Physics*, **79**, 184-186 (2003).

[25] H. Ohsato, A. Komura and S. Nishigaki, "Occupation in the Trigonal Columns of Tungsrenbronze-like $Ba_{6-3x}Sm_{8+2x}Ti_{18}O_{54}$ Solid Solutions," *J. Ceram. Soc. Jpn.*, **112[5]**, S1618-S1621 (2004).

[26] H. Ohsato, M. Mizuta, T. Ikoma, Z. Onogi, S. Nishigaki and T. Okuda, "Microwave Dielectric Properties of Tungstenbronze-Type $Ba_{6-3x}R_{8+2x}Ti_{18}O_{54}$ (R = La, Pr, Nd and Sm) Solid Solutions," *Ceram. Soc. Japan*, **106(2)**, 178-182 (1998); International edition, 106-185, 184-188 (1998).

CRYSTAL STRUCTURE ANALYSIS OF HOMOLOGOUS COMPOUNDS $ALa_4Ti_4O_{15}$ (A=Ba, Sr and Ca) AND THEIR MICROWAVE DIELECTRIC PROPERTIES

Y. Tohdo*, K. Kakimoto, H. Ohsato

Nagoya Institute of Technology, Gokiso-cho, Showa-ku, Nagoya 466-8555, Japan

T. Okawa, H. Okabe

Daiken Chemical Co., Ltd., 2-7-9 Hanaten-Nishi, Joto-ku, Osaka 536-0011, Japan

ABSTRACT

Relationships between the crystal structure and the microwave dielectric properties of homologous compounds $ALa_4Ti_4O_{15}$ (A=Ba, Sr and Ca) have been investigated for the design of new materials with higher dielectric constant (ε_r) and quality factor ($Q \cdot f$). X-ray diffraction techniques confirmed that $BaLa_4Ti_4O_{15}$ ceramics showed the smallest internal strain due to the ordering of Ba^{2+} and La^{3+} ions in A-sites. These ceramics showed a relatively high ε_r of 44.4. In contrast, $CaLa_4Ti_4O_{15}$ ceramics showed the highest $Q \cdot f$ of 50,246 (GHz) due to the decrease of distance between A-site ions and nearby oxygen layers. On the other hand, $SrLa_4Ti_4O_{15}$ ceramics showed a near zero temperature coefficient of resonant frequency τ_f of -8.4 (ppm/°C) that was lower than those measured for any of the other compounds of this study. Ionic displacement and crystallographic arrangement were discussed to clarify the effects of crystal structure on the microwave dielectric properties of $ALa_4Ti_4O_{15}$ (A=Ba, Sr and Ca) homologous compounds.

INTRODUCTION

Homologous compounds $ALa_4Ti_4O_{15}$ (A=Ba, Sr and Ca) in the AO-La_2O_3-TiO_2 ternary system (Fig. 1) are expected to be the materials of choice for base station resonator applications. Such application is required to have a high ε_r of more than 40 and a high $Q \cdot f$ of more than 30,000 (GHz). Some of the reported candidate materials with high ε_r and $Q \cdot f$ are listed such as $ZnNb_2O_8$-TiO_2: ε_r =37, $Q \cdot f$ =29,000 (GHz)[1] and BaO-TiO_2-ZnO: ε_r =36, $Q \cdot f$ =42,000 (GHz)[2]. In our previous studies[3-5], we have found $BaLa_4Ti_4O_{15}$ had a high ε_r of 44 and high $Q \cdot f$ of 47,000 (GHz) and also analyzed the crystal structures precisely based on the fundamental coordinates reported by Harre et al.[6] and Bontchev et al.[7] It was clarified that $ALa_4Ti_4O_{15}$ (A=Ba, Ca) had hexagonal perovskite-like layer structure and showed different ordering forms of A-site ions.

In the present study, we estimated internal strain of $ALa_4Ti_4O_{15}$ (A=Ba, Sr and Ca) ceramics and measured their microwave dielectric properties. We discuss the correlations between their crystal structures and microwave dielectric properties.

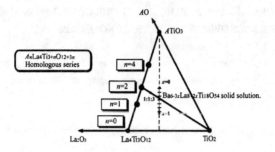

Fig. 1 AO-La_2O_3-TiO_2 (A=Ba, Sr and Ca) ternary system

EXPERIMENTAL PROCEDURE

Ceramic discs of $ALa_4Ti_4O_{15}$ (A=Ba, Sr and Ca) were prepared by a solid-state reaction method. Stoichiometric amounts of $BaCO_3$ (purity: 99.0%), $SrCO_3$ (98.0%), $CaCO_3$ (99.0%), La_2O_3 (99.9%) and TiO_2 (99.8%) powders were weighed and ball-milled in a polyethylene jar with zirconium balls and ethanol media for 24 h. The mixture was dried and calcined at 1000°C for 2 h in air. The calcined powders were ground and mixed with polyvinyl alcohol as a binder. The powders were screened through a mesh with 300μm openings and pressed under 98 (MPa) into discs. The sintering conditions in air were as follows; $BaLa_4Ti_4O_{15}$ (at 1600°C for 2 h), $SrLa_4Ti_4O_{15}$ (at 1550°C for 48 h) and $CaLa_4Ti_4O_{15}$ (at 1550°C for 24 h)[3,4].

The crystalline phase of sintered samples was identified by X-ray powder diffraction (XRPD). Furthermore, the internal strain was estimated from Full Width at Half-Maximum (FWHM) obtained by pattern decomposition method (Profit Ver. 3.00)[8]. The X-ray diffraction data for FWHM calculation was collected with a scanning step width of 0.020° and fixed counting time of 3 sec at each step. The six highlighted diffraction peaks shown in Fig. 2 were selected for the calculation. The relationship between FWHM (β) and internal strain (η) is shown in equation 1.

$$\beta=2\eta\tan\theta \qquad (1)$$

The dielectric constant (ε_r), unloaded Q values and temperature coefficients of the resonant frequency (τ_f) between 20 and 80°C were measured using a pair of parallel conducting Ag plates on the TE_{011} mode using Hakki and Coleman's method[9,10]. The apparent density was measured by Archimedes' method.

RESULTS AND DISCUSSION

Relative density and microwave dielectric properties of ALa$_4$Ti$_4$O$_{15}$ (A=Ba, Sr and Ca) ceramics are shown in Table 1. These ceramics samples exhibited good microwave dielectric properties. While BaLa$_4$Ti$_4$O$_{15}$ showed the highest ε_r of 44.4, CaLa$_4$Ti$_4$O$_{15}$ showed the highest $Q\cdot f$ of 50,246 (GHz). On the other hand, SrLa$_4$Ti$_4$O$_{15}$ showed a relatively good τ_f of -8.4 ppm/°C. SrTiO$_3$ structural units, which appear in the homologous structure of SrLa$_4$Ti$_4$O$_{15}$, have a large positive τ_f (1200 ppm/°C)[11], that seems to compensate for large negative τ_f of ALa$_4$Ti$_4$O$_{15}$ homologous compounds.

Table 1 Density and Microwave Dielectric Properties of ALa$_4$Ti$_4$O$_{15}$ (A=Ba, Sr and Ca)

Composition	d_r(%)	f(GHz)	ε_r	$Q\cdot f$(GHz)	τ_f(ppm/°C)
BaLa$_4$Ti$_4$O$_{15}$	98.4	6.9577	44.4	41008	-26.0
SrLa$_4$Ti$_4$O$_{15}$	98.9	6.8745	43.7	46220	-8.4
CaLa$_4$Ti$_4$O$_{15}$	94.8	6.9775	41.1	50246	-25.5

Powder X-ray diffraction patterns of the sintered ceramics are shown in Fig 2. Their crystalline phases were identified by comparing with *ICDD* cards No. 89-5557 (BaLa$_4$Ti$_4$O$_{15}$), No. 49-0254 (SrLa$_4$Ti$_4$O$_{15}$) and No. 36-1278 (CaLa$_4$Ti$_4$O$_{15}$).

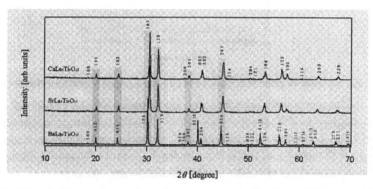

Fig. 2 Powder X-ray patterns of homologous compounds ALa$_4$Ti$_4$O$_{15}$ (A=Ba, Sr and Ca)

While CaLa$_4$Ti$_4$O$_{15}$ and SrLa$_4$Ti$_4$O$_{15}$ showed a similar XRPD patterns with space group $P\bar{3}m1$ (No. 164), BaLa$_4$Ti$_4$O$_{15}$ demonstrated a different pattern. X-ray diffraction peaks in the vicinity of 2θ=40° clearly indicate the differences in XRD characteristics among the samples. It is considered that BaLa$_4$Ti$_4$O$_{15}$ has a different space group $P\bar{3}c1$ (No.165). Figure 3 shows the FWHM of ALa$_4$Ti$_4$O$_{15}$ (A=Ba, Sr and Ca) as a function of tanθ. The calculated internal strain (η) is shown in Fig. 4.

149

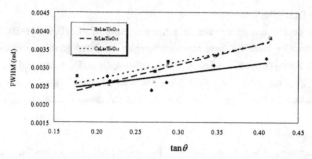

Fig. 3 FWHM of ALa$_4$Ti$_4$O$_{15}$ (A=Ba, Sr and Ca) homologous compounds

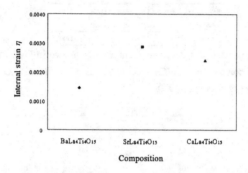

Fig. 4 Internal strain of ALa$_4$Ti$_4$O$_{15}$ (A=Ba, Sr and Ca) homologous compounds

The internal strain for BaLa$_4$Ti$_4$O$_{15}$ showed the lowest value of 1.45×10^{-3} among the three compounds. This strain was about half the value calculated for CaLa$_4$Ti$_4$O$_{15}$ and SrLa$_4$Ti$_4$O$_{15}$. As described in our previous paper[4], we found a rule of distribution for A-site ions of Ba^{2+} and Ca^{2+} depending on their ionic size in ALa$_4$Ti$_4$O$_{15}$ (A=Ba, Ca) crystals. Figure 5 shows a schematic representation of ALa$_4$Ti$_4$O$_{15}$ (A=Ba, Ca) homologous structures.

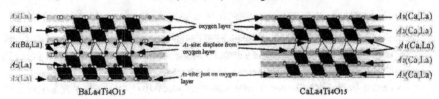

Fig. 5 Schematic representation of ALa$_4$Ti$_4$O$_{15}$ (A=Ba, Ca) crystal structure
as viewed along (2$\bar{1}$0) plane

A_1-sites locate near the empty octahedron space, but other A_2- and A_3-sites locate near the TiO_6 octahedron. It indicates that the free space for A_1-sites is estimated to be larger than those of A_2- and A_3-sites. The ionic radius of Ba^{2+} (r=1.61 Å: 12-coordination) is larger than that of La^{3+} (r=1.36 Å: 12-coordination); therefore, Ba^{2+} predominantly occupies A_1-sites and La^{3+} occupies each A-sites (A_1-A_3). In the case of $CaLa_4Ti_4O_{15}$, however Ca^{2+} and La^{3+} occupy all A-sites, because both Ca^{2+} (r=1.34 Å: 12-coordination) and La^{3+} (r=1.36 Å) ions have nearly the same ionic size.

Figure 6 shows displacement of A-site ions (A=Ba, Ca and La) from oxygen packing layers and that of B-site ion (B=Ti) from the center of octahedrons for $ALa_4Ti_4O_{15}$ (A=Ba, Ca). Figure 7 shows the scheme of A-site arrangement for $ALa_4Ti_4O_{15}$ (A=Ba, Sr and Ca) to grasp these differences easily.

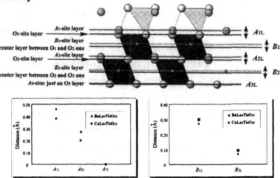

Fig. 6 Distance of A-site ions (A=Ba, Ca and La) from oxygen packing layer and B-site
ions (B=Ti) from the center of octahedron

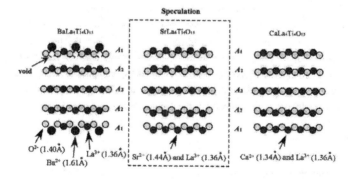

Fig. 7 Schematic representation of behavior for A-site ions (A=Ba, Sr, Ca and La)

According to Fig. 6, all of A- and B-site cations (A=Ba, La and B=Ti) of $BaLa_4Ti_4O_{15}$ displace with larger distances from oxygen layers and from the center of TiO_6 octahedron than those (A=Ca, La and B=Ti) of $CaLa_4Ti_4O_{15}$. Especially, Ba^{2+} of A_1-site makes a move from O_1-site layer significantly. This seems to be because the ionic radius of Ba^{2+} (r=1.61 Å) is larger than that of O^{2-} (r=1.40 Å: 6-coordination). In the case of $CaLa_4Ti_4O_{15}$, ionic radius of Ca^{2+} (r=1.34 Å) and La^{3+} (r=1.36 Å) are close to that of O^{2-} (r=1.40 Å). Therefore, these ions were aligned without large displacement along the oxygen layer (Fig. 7). The crystal structure of $SrLa_4Ti_4O_{15}$ is presumed to be similar to that of $CaLa_4Ti_4O_{15}$, because the ionic radius of Sr^{2+} (r=1.44 Å: 12-coordination) is also close to that of O^{2-} (r=1.40 Å).

$BaLa_4Ti_4O_{15}$ showed the smallest internal strain of all specimens because Ba^{2+} and La^{3+} ions are well ordered at A-sites. In the case of $CaLa_4Ti_4O_{15}$, Ca^{2+} and La^{3+} occupy all A-sites without ordering. As a result, the lattice plane spacing of A-sites ions showed irregularity, which resulted in large internal strain compared with $BaLa_4Ti_4O_{15}$. On the other hand, the internal strain of $SrLa_4Ti_4O_{15}$ almost equals that of $CaLa_4Ti_4O_{15}$. In the case of $SrLa_4Ti_4O_{15}$, therefore it can be guessed that Sr^{2+} (r=1.44 Å) and La^{3+} (r=1.36 Å) ions probably occupy all A-sites randomly similar to the case for $CaLa_4Ti_4O_{15}$. Since the cations of $BaLa_4Ti_4O_{15}$ demonstrated larger displacement, the ionic polarization greater than that of $CaLa_4Ti_4O_{15}$ and $SrLa_4Ti_4O_{15}$. Accordingly, $BaLa_4Ti_4O_{15}$ showed the highest ε_r of three compounds. In contrast, $CaLa_4Ti_4O_{15}$ showed the highest $Q \cdot f$. Ca^{2+} and La^{3+} ions were filled up along the oxygen layer and it caused a decrease of the distance between A-sites ions and the adjacent oxygen layer. Since oscillation of the Ca^{2+} and La^{3+} ions of $CaLa_4Ti_4O_{15}$ was inhibited, dielectric loss (tanδ) was small and the highest $Q \cdot f$ was measured.

CONCLUSION

It was clarified that internal strains of $ALa_4Ti_4O_{15}$ (A=Ba, Sr and Ca) homologous ceramics had a close correlation with the ordering of A-sites ions. Internal strain of $BaLa_4Ti_4O_{15}$ showed the lowest value due to the ordering of A-site ions (A=Ba, La). The relationships between the displacement of A- and B-sites ions from oxygen layers and the center of near by TiO_6 octahedra and microwave dielectric properties of these compounds were discussed. While $BaLa_4Ti_4O_{15}$ ceramics showed the highest ε_r of 44.4, $CaLa_4Ti_4O_{15}$ ceramics showed the highest $Q \cdot f$ of 50,246 (GHz) due to the decrease of the distance between A-sites ions and adjacent oxygen layers.

ACKNOWLEDGEMENTS

This work was supported by a grant from the NITECH 21st Century COE Program "World Ceramics Center for Environmental Harmony". The authors would like to thank Mr. A.

Harada, President of Daiken Chemical Co., Ltd. for his arrangement in the measurement of microwave dielectric properties.

REFERENCES

[1]D-W. Kim, K.H. Ko, D-K. Kwon and K. S. Hong, "Origin of microwave dielectric loss in $ZnNb_2O_6$-TiO_2", *J. Am. Ceram. Soc.*, **85,** 1169-1172 (2002).

[2]W. Shunhua, S. Hao, Z. Yushuans and W. Guoqing, "BaO-TiO_2 microwave ceramics", *J. Eur. Ceram. Soc.,* **23,** 2565-2568 (2002).

[3]T. Okawa, K. Kiuchi, H. Okabe and H. Ohsato, H., "Microwave dielectric properties of $Ba_nLa_4Ti_{3+n}O_{12+3n}$ homologous compounds and substitution of trivalent cations for La", *Ferroelectrics,* **272,** 345-350 (2002).

[4]Y. Tohdo, T. Okawa, H. Okabe, K. Kakimoto and H. Ohsato, "Microwave dielectric homologous materials $ALa_4Ti_4O_{15}$ (A=Ba, Ca, Sr) with high $Q \cdot$ high dielectric constant for base station", *Key Eng. Mater.,* **269,** 203-206 (2004).

[5]Y. Tohdo, T. Okawa, H. Okabe, K. Kakimoto and H. Ohsato, "Microwave dielectric properties and crystal structure of homologous compounds $ALa_4Ti_4O_{15}$ (A=Ba, Ca and Sr) for base station", *J. Eur. Ceram. Soc.,* accepted for publication.

[6]N. Harre, D. Mercurio, G. Trolliard and B. Frit, "Crystal structure of $BaLa_4Ti_4O_{15}$, member n=5 of the homologous series $(Ba, La)_nTi_{n-1}O_{3n}$ of cation-deficient perovskite-related compounds", *Mater. Res. Bull.,* **33,** 1537-1548 (1998).

[7]R. Bontchev, F. Weill and J. Darriet, "New 5H hexagonal perovskite-like oxides (The series $ALa_4Ti_3RuO_{15}$, where A=Ca, Sr, Ba)", *Mater. Res. Bull.,* **27,** 931-938 (1992).

[8]H. Toraya, "Whole-powder-pattern fitting without reference to a structural model: application to X-ray powder diffraction data", *J. Appl. Cryst.,* **19,** 440-447 (1986).

[9]B. W. Hakki and P. D.Coleman, "A dielectric resonator method of measuring inductive in the millimeter range", *IRE Trans. Microwave Theory & Tech.,* **MTT-8,** 402-410 (1960).

[10]Y. Kobayashi and M. Katoh, "Microwave measurement of dielectric properties of low-loss materials by the dielectric rod resonator method", *IEEE Trans. Microwave Theory & Tech.,* **33** 586-592 (1985).

[11]P. L. Wise, I. M. Reaney, W. E. Lee, T. J. Price, D. M. Iddles and D. S. Cannell, "Structure-microwave property relations in $(Sr_xCa_{(1-x)})_{n+1}Ti_nO_{3n+1}$", *J. Eur. Ceram. Soc.,* **21,** 1723-1726 (2001).

153

EFFECTS OF IONIC RADII AND POLARIZABILITY ON THE MICROWAVE DIELECTRIC PROPERTIES OF FORSTERITE SOLID SOLUTIONS

T. Sugiyama, H. Ohsato, T. Tsunooka, K. Kakimoto
Nagoya Institute of Technology, Gokiso-cho, showa-ku, Nagoya 466-8555, Japan

ABSTRACT

Silicates composed of SiO_4 tetrahedroa which have a half covalent bonds are expected low dielectric constants and hopeful as millimeter-wave dielectric materials. It was reported forsterite (Mg_2SiO_4) with very high quality factor (240000GHz) was prepared by adopting very pure MgO and SiO_2 powder.[1] But the temperature coefficient of resonant frequency (τ_f) of forsterite showed -70ppm/°C. We succeeded to improve τ_f 0ppm/°C by addition of Rutile (TiO_2), but the $Q \cdot f$ decreased.[2] In this study, we tried to improve τ_f of forsterite by forming substituting Ni and Mn for Mg. The microwave dielectric properties were affected by the ionic radii and polarizability. Ions with larger ionic radii and polarizability bring τ_f toward minus, therefore, we will report the effects of ion with the smaller ionic radii and polarizabirity than Mg.

INTRODUCTION

Recently, microwave telecommunication has been developed for various applications. The increase in the amount of information to be transported needs millimeter-wave telecommunication, which can transmit a large amount of information at a very high speed. This is important in intelligent transport systems (ITS), ultra high speed wireless LAN and satellite broadcasting. The important characteristics required for a dielectric material used in millimeter-wave telecommunication systems are as follows. (a) high quality factor ($Q \cdot f$) to achieve high selectivity, (b) low dielectric constant (ε_r) to reduce the delay time of electronic signal transmission and (c) nearly zero temperature coefficient of resonant frequency (τ_f) for frequency stability. Silicates build on silica tetrahedron with about 55% of covalent bonding. Hence silicates are suitable dielectric materials for millimeter-wave r.εcommunication due to their low ε_r.[1,2,3,4,5]

We have studied dielectric properties of forsterite solid solutions ($Mg_{2-x}M_xSiO_4$, M=Ca, Mn). The crystal structure of forsterite solid solution is olivine, which is composed of M1, M2 and Si-site as shown Fig.1.[6] In the series, it is supposed τ_f is concerned with ionic radii in M1 and M2-site. In order to confirm the assumption, we investigate the dielectric properties of $Mg_{2-x}Ni_xSiO_4$ solid solutions and the influence of the substitution of Mg by Ni.

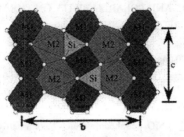

Fig. 1 crystal structure of olivine silicate displayed as Pbnm

EXPERIMENTAL PROCEDURE

High purity chemicals such as MgO (xx%), NiO (99.9%) and SiO2 (99.9%) powders were weighed and mixed in stoichiometric ratios then ball milled for 24 h using ZrO2 balls. After drying the powder for $Mg_{2-x}Ni_xSiO_4$ ceramics was calcined in the condition as shown in table 1. Calcining temperature and time were set to prepare the object. To prepare the composition x=2.0 twice calcination was needed. The calcined powder was ball milled again for 24h and dried. The powder was pressed into cylindrical shape under a uni-axial pressure of 7.84MPa and CIP of 200MPa. The pellets were then sintered at 1400℃ for 2h in air.

Table I Condition of Calcination

composition x	0.05	0.5	1.0	1.5	2.0
temperature	1150℃	1220℃	1240℃	1260℃	1350℃
time	2h	6h	10h	12h	12h

The crystalline phases of the samples were identified by powder X-ray diffraction. Lattice parameters are refined using the computer program WPPF for whole-powder-pattern fitting.[7] Microwave dielectric properties were measured by Hakki and Colemans' method[8] using the TE_{011} mode with a network analyzer. Volume of Ni, Mg and Mn site were calculated with computer program Cygnus. Date of crystal structure for the calculation was quoted from papers.[6,9,10]

RESULTS AND DISCCUSION

In the composition x=0.05, 0.5, 1.0, 1.5 and 2.0, all the peaks of XRD were confirmed to the peaks of $Mg_{2-x}Ni_xSiO_4$ solid solutions and secondary phase were not observed. The refined lattice parameters of $Mg_{2-x}Ni_xSiO_4$ were shown in Fig. 2. In all axes the parameters decreased with increase of x. The ionic radii of 6 coordinated Ni^{2+} is smaller than Mg^{2+}, therefore lattice parameters were shortened. From the results, it was confirmed $Mg_{2-x}Ni_xSiO_4$ solid solutions were prepared.

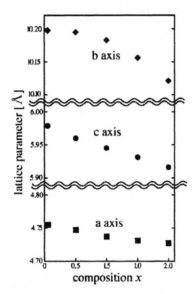

Fig. 2 lattice parameter of $Mg_{2-x}Ni_xSiO_4$ as a function of composition x

τ_f of $Mg_{2-x}Ni_xSiO_4$ were shown in Fig. 3 as a function of composition x. τ_f shited to 0ppm/°C linearly with the increase of amount of Ni. It was supposed τf was closely connected with ionic radii of M1-site ions.

τ_f of M_2SiO_4 corresponded to the ionic radii of M was shown in Fig. 4.[11] The smaller ionic radii of cation resulted in a τf closer to 0ppm/c τf was. As shown in Fig. 5, in the ideal HCP model olivine structure, oxygen was located on the same plane as M1 along to [010]. However in the actual olivine structure the plane oxygen were located on was deviated from the plane M1 was filled in. The deviated distance L, free space (FS) of M1 and M2-site and free space rate (FSR) were shown in Table 2. FS and FSR were defined and calculated as shown in Fig. 6.

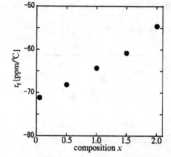

Fig. 3 tf of $Mg_{2-x}Ni_xSiO_4$ as
a function of composition x

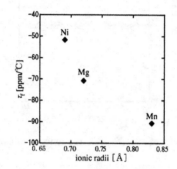

Fig. 4 τ_f of M_2SiO_4 (M=Ni, Mg,
Mn) as a function of ionic radii

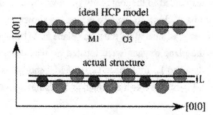

Fig. 5 Deviation of plane oxygen located on from plane M1 filled in along [010] direction

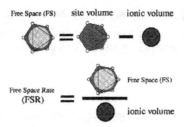

Free Space (FS)   site volume   ionic volume

Free Space Rate
(FSR)   = $\dfrac{\text{Free Space (FS)}}{\text{ionic volume}}$

Fig. 6 Definition of Free Space (FS) and Free
Space Rate (FSR)

Table II L, FS and FSR of M_2SiO_4

composition	τ_f [ppm/°C]	L [Å]	FS of M1 [Å³]	FS of M2 [Å³]	FSR of M1	FSR of M2
Ni_2SiO_4	-54.6	0.18	10.2	10.6	7.41	7.69
Mg_2SiO_4	-70.9	0.20	10.2	10.9	6.52	6.96
Mn_2SiO_4	-90.6	0.29	10.9	11.9	4.54	4.96

The larger ionic radii were, the longer deviant distance of L. The long L indicated large distortion from HCP. τ_f was shifted toward more negative value with increase of L. From the results, it was supposed a composition with HCP of oxygen was expected for a superior τ_f.

Dielectric properties of $Mg_{2-x}Ni_xSiO_4$ solid solutions were shown in Table 3. Q·f decreased with the increase of x. Relative density of these samples was low, therefore, Q·f of them showed low value. Improvement of Q·f is expected by optimizing the preparing process.

Polarizability of Ni^{2+} is smaller than Mg^{2+}, therefore ε_r was considered to become low. However ε_r of $Mg_{2-x}Ni_xSiO_4$ solid solutions indicated the trend to increase with x. FS of M2-site of Ni_2SiO_4 was smaller than that of Mg_2SiO_4. However FSR of M1 and M2-site of Ni_2SiO_4 was larger than that of Mg_2SiO_4. Therefore, Ni was considered easier to move than Mg and ε_r of Ni2SiO4 become larger than Mg2SiO4.

Table 3 Dielectric properties of Ni_2SiO_4

composition x	ε_r	Q·f [GHz]	τ_f [ppm/°C]	relative density [%]
0.05	6.61	121121	-71.2	94.8
0.5	6.49	45721	-68.1	93.0
1.0	6.61	26577	-64.3	91.1
1.5	6.91	25103	-60.8	95.9
2.0	7.04	22271	-54.5	95.7

CONCLUSION

In M_2SiO_4 olivine silicates adopting smaller cation as M shorten the deviant distance between the plane oxygen filled in and the plane M filled in, consequently the oxygen array got close to HCP. It was considered the structure close to HCP produced a τf close to 0ppm/°c.

REFERENCE

[1] Tsunooka, T.; Androu, M.; Higashida, Y.; Sugiura, H.; Ohsato, H. "Effects of TiO$_2$ on sinterability and dielectric properties of high-Q forsterite ceramics. Journal of the European Ceramic Society, 23(14), 2573-2578 (2003)

[2] Tsunooka, Tsutomu; Sugiyama, Tomonori; Ohsato, Hitoshi; Kakimoto, Ken-ichi; Andou, Minato; Higashida, Yutaka; Sugiura, Hirotsugu. Development of forsterite with high Q and zero temperature coefficient *auf for Millimeterwave Dielectric Ceramics. Key Engineering Materials, 269, 199-202 (2004)

[3] Ohsato, H., Tsunooka, T., Ando, M., Ohishi, Y., Miyauchi, Y., Kakimoto, K., "Millimeter-wave dielectric ceramics of alumina and forsterite with high quality factor and low dielectric constant". Journal of the Korean Ceramics Society, 40(4), 350-353 (2003)

[4] T. Tsunooka, H. Ohsato, T. Sugiyama, "Manufacture of ceramics for planar high-frequency circuit", PCT Int. Appl, 33 pp (2004)

[5] Y. Ohishi, Y. Miyauchi, H. Ohsato, K. Kakimoto, "Controlled Temperature Coefficient of resonant frequency of Al$_2$O$_3$-TiO$_2$ ceramics by annealing Treatment", Japanese Jor. Appl. Phys, Part2: Letters & Express Letters, 43 (6A), 749-51 (2004)

[6] R. Hanzen, "Effects of temperature and pressure on the crystal structure of forusterite," American Mineralogist, 61,1280-93 (1976)

[7] H. Toraya, "Whole-Powder-Pattern Fitting Without Reference to a structural Model: Application to X-ray Powder Diffractometer Data", Journal of Applied Crystallography, 19, 440-447 (1986)

[8] W. Hakki, D. Coleman, IRE Trans. Microwave Theory & Tech., MTT-8, 402 (1960)

[9] O. Tamada, K. Fujino, S. Sasaki, "Structures and Electron distributions of α-Co$_2$SiO$_4$ and Ni$_2$SiO$_4$ (Olivine structure)", Acta Cryst. B39, 692-7 (1983)

[10] S. Yamazaki, H. Toraya, "Rietveld refinement of site-occupancy parameters of Mg$_2$-xMnxSiO$_4$ using a new weight function in least-squares fitting", J. Appl. Cryst., 32, 51-59 (1999)

[11] D. Shannon, "Revised effective radii and systematic of interatomic distances in halides and chalcogenides", Acta Cryst, A32, 751-767 (1976)

MICROWAVE CHARACTERISATION OF CALCIUM FLUORIDE IN THE TEMPERATURE RANGE 15-300K

Mohan V. Jacob
Electrical and Computer Engineering, James Cook University, Townsville, Australia

Janina Mazierska
Institute of Information Sciences, Massey University, Palmerston North, New Zealand
and
Electrical and Computer Engineering, James Cook University, Townsville, Australia

J. Krupka
Instytut Mikroelektroniki i Optoelektroniki Politechniki Warszawskiej,
Koszykowa 75, 00-662 Warszawa, Poland.

ABSTRACT
    Calcium Fluoride crystals are extensively used in many optical applications. In order to test the feasibility of using CaF_2 crystal at microwave frequencies, a rod shaped CaF_2 crystal is characterised using two types of dielectric resonators; namely Hakki-Coleman dielectric resonator and dielectric post resonator. The precise measurement results of the perpendicular component of complex permittivity of CaF_2 as a function of temperature (from 15 K to 300 K) at frequencies 17 GHz and 29 GHz is presented in this paper. We have used the multifrequency S-parameter measurement technique and the Transmission Mode Q-Factor technique to improve the accuracy of measurements and hence precise values of permittivity and loss tangent. The typically used microwave parameters such as $Q_0 \times f_0$, temperature coefficient of frequency and temperature coefficient of permittivity are also analysed for both measurement techniques. The $Q_0 \times f_0$ and temperature coefficient of frequency depends on the measurement technique where as the temperature coefficient of permittivity remains independent of measurement technique. This implies that temperature coefficient of permittivity data represents a material property.

INTRODUCTION
    The recent advancement of wireless communication systems demands dielectric materials with low losses. Also devices fabricated using High Temperature Superconducting (HTS) materials need low loss dielectric materials to use in conjunction with superconductors. In order to design a circuit or device it is essential to know the characteristics of all materials, which will be used for fabrication. At microwave frequencies the device performance is often simulated from the two parameters, real part of permittivity and loss tangent.
    Calcium Fluoride (CaF_2) crystals are used in many optical applications, including mirror substrates for UV laser systems, windows, lenses and prisms for ultraviolet, visible and infrared frequencies[1-4]. CaF_2 is highly resistant to the environmental changes. Low solubility and wide transmission makes CaF_2 practical for many applications, including mirror substrates for UV laser systems, windows, lenses and prisms for UV and IR applications. CaF_2 can be used in microwave planar circuits as a substrate material due to its low relative permittivity and low losses. Another possibility could be a hybrid HTS - Silicon technology for microwave circuits. Due to the low

relative permittivity and low losses, Calcium Fluoride can find applications in microwave planar circuits as a substrate material. Another possibility could be a hybrid High Temperature Superconductor (HTS)- Silicon technology for microwave circuits as investigated in the following references[2,3].

CaF$_2$ is grown by the Stockbarger technique or the Brigdman method in diameter up to about 200 mm. Calcium Fluoride (VUV grade) crystals have the transmission range from 0.19 μm to 7.2 μm and low refractive index from about 1.35 to 1.51 through this range[1]. IR grade Calcium Fluoride is transparent up to 12 μm.

The complex permittivity data of CaF$_2$ are available at cryogenic temperatures and at microwave frequencies[4-6]. But none of these papers are complete with all the microwave parameters such as permittivity, loss tangent, $Q_0 \times f_0$, temperature coefficient of frequency (τ_f) and temperature coefficient of permittivity (τ_e). In this paper we have used two types of measurement fixtures to analyse all the microwave parameters of the CaF$_2$ crystal[7].

EXPERIMENTAL TECHNIQUES

The dielectric resonator technique is typically used for the microwave characterisation of dielectric materials. Among the different types of dielectric resonators, Hakki-Coleman dielectric resonator (HCDR) and dielectric post resonators (DPR) are commonly used to characterise rod shaped dielectric materials at microwave frequencies[8-12]. We have used a TE$_{01\delta}$ mode dielectric post resonator and a TE$_{011}$ mode Hakki-Coleman dielectric resonator to characterise CaF$_2$ crystal[10-12] as shown in Fig. 1.

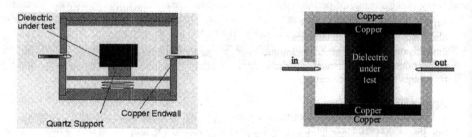

Fig. 1 The schematic diagram of dielectric post and Hakki-Coleman dielectric resonators

The experimental system used for the microwave characterisation consists of Network Analyser (HP 8722C), closed cycle refrigerator (APD DE-204), temperature controller (LTC-10), vacuum Dewar, a PC and the dielectric resonator (either HCDR or DPR). The resonator containing the CaF$_2$ sample was cooled from room temperature to approximately 13 K. The TE$_{011}$ mode resonance is identified at frequency of around 29 GHz for HCDR and 17.2 GHz for DPR. The S$_{21}$, S$_{11}$ and S$_{22}$ parameter data around the resonance were measured at the lowest temperature. Transmission Mode Q-Factor technique[13] has been used to eliminate the parasitic losses such as influence of phase shift, cross talk and to precisely compute coupling coefficients k$_1$ and k$_2$. The unloaded Q-factor is calculated using[14]:

$$Q_0(T) = Q_L(T)[1 + k_1(T) + k_2(T)]$$ (1),

where Q_L, is the loaded Q-factor obtained from the S_{21} measurements.

The S_{21} parameter is measured as a function of temperature from 15 K to 300 K. The coupling coefficient for each measurement temperature is calculated using a simplified TMQF[15] and hence the unloaded Q-factor. The perpendicular component of the real part of relative permittivity and loss tangent (tanδ) of CaF_2 was computed from the measured resonant frequency and unloaded Q-factor respectively for different temperatures.

RESULTS AND DISCUSSIONS

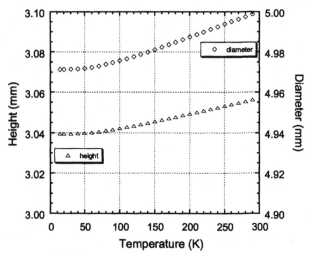

Fig. 2 The dimensions of the CaF_2 crystal at various measurement temperatures.

A Calcium Fluoride crystal has been characterised as a function of temperature at microwave frequencies. We have calculated the dimensions of the crystal at low temperatures using the of linear thermal expansion coefficient[16] to increase the accuracy of measurement. Figure 2 shows the estimated dimensions at various temperatures. The resonant frequency of both the resonators are shown in Fig. 3 for the temperature range 15 K to 300 K. The unloaded Q-factor of the resonators is calculated using the TMQF method from the S-parameter measurement data. RF design engineers are interested in the product of unloaded Q-factor and resonant frequency, which is shown in Fig. 4. Figure 4 indicates that the $Q_0 \times f_0$ is not a constant for a given temperature as expected. Even though it could give an indication of how good is the material for a given application, it is more dependent

163

on the type of resonator.

We have calculated the loss of the material at microwave frequencies using both techniques. The results shown in Fig. 5 for the HC resonator is at a frequency of 29 GHz and for the post resonator is at 17.2 GHz. Below 100 K, the loss tangent values are similar for both techniques. This may attribute to the uncertainty in measurements especially at very low temperatures. For HCDR the influence of the cavity walls are higher than that of DPR. This will contribute higher error in measurement. The uncertainty in loss tangent measurements of the DPR is much smaller compared to HCDR for material of moderate losses like CaF_2. The loss tangent varies between 6×10^{-5} and 23×10^{-5} (for DPR – 17.2 GHz) and 7×10^{-5} and 32×10^{-5} (for HCDR – 29 GHz) in the temperature range 15 K to 300 K.

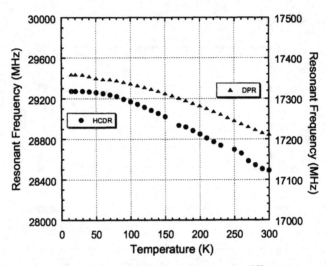

Fig. 3 The resonant frequencies of the CaF_2 resonators at different temperatures.

The calculated values of the perpendicular component of the permittivity are shown in Fig. 6. The difference in the values of permittivity of CaF_2 obtained using HCDR and DPR is between 0.7% to 2.5% in the measured range of temperature. This could be due to the uncertainty in the dimensions. As shown in Fig. 1, the sample is sandwiched between two Cu plates in the case of HCDR and could have resulted in this higher deviation at higher temperatures.

The temperature coefficient of frequency and permittivity are estimated and shown in Figures 7 and 8. The deviation in the values of τ_f is significantly higher (more than 100%) where as the value of τ_ε is similar in both cases. This indicates that the more reliable parameter that a design engineer could consider is the temperature coefficient of permittivity and not temperature coefficient of frequency.

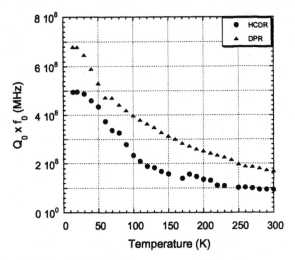

Fig. 4 The $Q_0 \times f_0$ of the CaF_2 resonators at different temperatures.

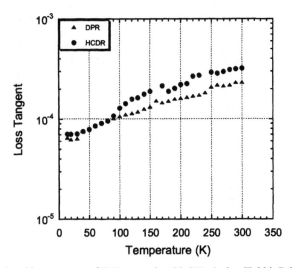

Fig. 5 The calculated loss tangent of CaF_2 crystal at 29 GHz (using Hakki-Coleman dielectric resonator) and at 17.2 GHz (using the dielectric post resonator).

165

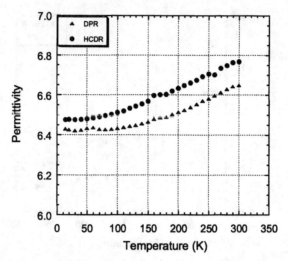

Fig. 6 The calculated perpendicular component of real part of permittivity of CaF$_2$ crystal at 29 GHz (HCDR) and at 17.2 GHz (DPR).

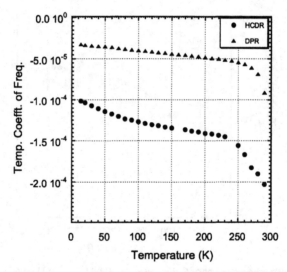

Fig. 7 The estimated temperature coefficient of frequency using the HCDR and DPR techniques.

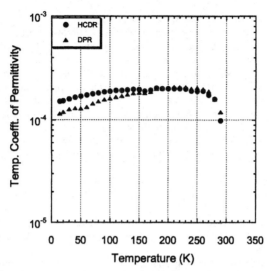

Fig. 8 The estimated temperature coefficient of permittivity using the HCDR and DPR techniques.

CONCLUSIONS

The complex permittivity of Calcium Fluoride is measured using a Hakki-Coleman dielectric resonator and a dielectric post resonator at frequencies of 29 GHz and 17.2 GHz respectively in the temperature range of 15 K to 300 K. The permittivity values varies between 6.43 and 6.65 at frequency of 17.2 GHz and 6.5 and 6.76 at frequency of 29 GHz. The corresponding loss tangent values are 6×10^{-5} and 23×10^{-5} and 7×10^{-5} and 32×10^{-5} in the temperature range of 14 K to 290 K. The calculations of $Q_0 \times f_0$ and temperature coefficient of frequency show that these two parameters not just reveal the material properties but also depend on the measurement technique. On the other hand the temperature coefficient of permittivity reflects on the material property irrespective of the measurement techniques.

ACKNOWLEDGEMENTS

This work was done under the financial support of ARC Discovery Project DP0449996. MVJ also acknowledges the ARC Australian Research Fellowship.

REFERENCES

[1] [online]: Del Mar Ventures, www.sciner.com/Opticsland/CaF2.htm

[2] O.W. Kwon, B.W. Langley, R.F.W. Pease, M.R. Beasley, *IEEE Electron Devices Lett.* **EDL-8**, 582 (1987)

[3] S.S. Bhagwat, A.R. Bhangale, J.M. Patil, V.S. Shirodkar, R. Pinto, P.R. Apte and S.P. Pai, "Growth of CaF2 buffer on Si using low energy cluster beam deposition technique and study of its

properties", *Brazilian Journal of Physics*, **29** No. 2 (1999)

[4] M. V. Jacob, J. Mazierska, D. Ledenyov and J. Krupka "Microwave Characterisation of CaF_2 at cryogenic temperatures using Dielectric Resonator Technique", *European Journal of Ceramic Society*, **23** 2617-2622 (2003).

[5] J. G. Hartnett, A. Fowler, M. E. Tobar and J. Krupka, "Dielectric properties of single crystal fluorides at microwave frequencies and cryogenic temperatures" *private communication* (2004).

[6] R. G. Geyer, J. Baker-Jarvis and J. Krupka, "Dielectric characterisation of single crystal LiF, CaF_2, MgF_2, BaF_2 and SrF_2 at microwave frequencies" *private communication* (2004).

[7] The Institute of Materials Research and Technology, Warsaw, Poland.

[8] Y. Kobayashi, M. Katoh, "Microwave measurement of dielectric properties of low-loss materials by the dielectric rod resonator method", *IEEE Transactions on Microwave Theory and Techniques* **33**, 586-592 (1985).

[9] L. Chen, C. K. Ong and B. T. G. Tan, "Amendment of cavity perturbation method for permittivity measurement of extremely low-loss dielectrics, *IEEE Transactions on Instrumentation and Measurement*, **48**, 1031-36 (1999).

[10] J. Krupka, K. Derzakowski, M.E. Tobar, J. Hartnett, and R.G. Geyer, "Complex permittivity of some ultralow loss dielectric crystals at cryogenic temperatures", *Measurement Science and Technology*, **10**, 387-392 (1999).

[11] M. V. Jacob, J. Mazierska, K. Leong and J. Krupka: "Microwave Properties of Low Loss Polymers at Cryogenic Temperatures" *IEEE Transactions on Microwave Theory and Techniques*, **50**, 474-480, (2002).

[12] C Zuccaro, M Winter, N Klein and K Urban *Journal of Applied Physics* **82**, 5695-5704 (1997).

[13] K. Leong and J. Mazierska, "Precise Measurements of the Q-factor of Transmission Mode Dielectric Resonators: Accounting for Noise, Crosstalk, Coupling Loss and Reactance, and Uncalibrated Transmission Lines", *IEEE Tran. on MTT* , **50**, 2115-2127 (2002).

[14] E. L. Ginzton, "Microwave Measurements", *McGraw Hill Book Co., New York* (1957).

[15] M. V. Jacob, J. Mazierska, K. Leong and J. Krupka, "Simplified method for measurements and calculations of coupling coefficients and Q_o-factor of high temperature superconducting dielectric resonators", *IEEE Trans of Microwave Theory and Techniques*, **49**, 2401-2407, (2001).

[16] Y. S. Touloikian, R.K. Kirby, R.E. Taylor and T.Y.R. Lee, "Thermal expansion - Nonmetallic solids", *Thermophysical Properties of Matter Data Series*, **13**, IFI/Plenum, New York, (1977).

HIGH-QUALITY 2 INCH $La_3Ga_{5.5}Ta_{0.5}O_{14}$ AND $Ca_3TaGa_3Si_2O_{14}$ CRYSTALS FOR OSCILLATORS AND RESONATORS

Christine F. Klemenz, Jun Luo, Dhaval Shah,
Advanced Materials Processing and Analysis Center (AMPAC)
University of Central Florida
Orlando, FL, 32816

ABSTRACT

$Ca_3Ga_2Ge_4O_{14}$ (CGG)-type crystals have attracted much attention over the past years for their outstanding piezoelectric properties, superior to quartz, offering new possibilities in bulk (BAW) and surface acoustic wave (SAW) devices. Among them, most investigated are the isomorphs $La_3Ga_5SiO_{14}$ (LGS), $La_3Ga_{5.5}Nb_{0.5}O_{14}$ (LGN), and $La_3Ga_{5.5}Ta_{0.5}O_{14}$ (LGT), so-called langasites, commonly denoted herein LGX (X = S, N, and T). Recently, the quaternary CGG isomorphs, $Ca_3TaGa_3Si_2O_{14}$ (CTGS) and $Ca_3NbGa_3Si_2O_{14}$ (CNGS) became of interest as they have an ordered crystal structure that may lead to a better consistency in acoustic properties and a higher Q than LGT. Today, the main limitation towards commercialization of any of these new piezoelectrics is the insufficient homogeneity of the crystals, which lead to non-reliable materials parameters measurements and a spread in device characteristics.

We report on the growth of high-quality large-diameter 2" LGT and 1.5" CTGS crystals. The study of inherent and growth-induced crystal defects combined with systematic Czochralski growth process optimization have led to these colorless transparent crystals. The high-quality of the LGT crystals was confirmed by X-ray synchrotron topography.

The LGT crystals were processed into Y-cut plano-convex resonators by MtronPTI (Orlando). The $Q \cdot f$ product was measured from fundamental to 9th overtone. Values exceeding the limit of AT- and SC-cut quartz were obtained.

INTRODUCTION

$La_3Ga_5SiO_{14}$ (LGS), $La_3Ga_{5.5}Nb_{0.5}O_{14}$ (LGN), $La_3Ga_{5.5}Ta_{0.5}O_{14}$ (LGT) and $Ca_3TaGa_3Si_2O_{14}$ (CTGS) belong to the $Ca_3Ga_2Ge_4O_{14}$ (CGG) structure-type, with space group P321 and the same trigonal crystal class 32 as quartz. Czochralski growth of LGS, LGN, and LGT, commonly denoted herein LGX (X=S, N, and T), has been very actively pursued for more than a decade, driven by the potential of these materials to replace quartz in a new generation of oscillators and resonators.

The advantages of LGX over quartz include higher electromechanical coupling, which allows wider bandwidth and enables devices to be made smaller, and a higher Q value (or lower acoustic loss)[1,2] which is important for low phase noise and enables higher-frequency operation. Many isomorphs of CGG, including LGT and CTGS, have in common with quartz the existence of shear-wave velocities that are, to first order, independent of temperature for certain crystal orientations. In addition, it has been postulated that LGX has a lower acceleration sensitivity[3,4] than quartz. These properties may lead to significant improvements in frequency control applications (communication, radar, identification, and navigation systems)[5]. Unlike quartz, these materials have no phase transition below their melting point ($\approx$ 1500°C for LGT). This allows higher-temperature device processing, and high-temperature applications. Due to their congruent melting behavior, LGX, CTGS, and CNGS crystals can be pulled directly from the melt by Czochralski.

Despite their huge potential, these materials have not so far obtained broad commercial acceptance: inhomogeneities in typical crystals lead to a wide scatter in acoustic properties and associated device characteristics[6,7,8,9]. It is believed that difficulties in growing homogeneous LGX are associated with the crystals being intrinsically disordered. In LGX, certain lattice sites are irregularly populated by more than one element, leading to fluctuations in crystal composition. In the LGT crystal structure, the octahedral site can be occupied by either Ta or Ga. It has also been postulated that this disorder causes random distortions of the crystal lattice leading to a lower Q value and electromechanical coupling than achievable. Thus, quaternary CGG isomorphs with ordered structure were investigated[10]. Among them, CNGS and CTGS have temperature-compensated orientations and were studied and grown by a few groups[11,12,13].

Despite its obvious importance, only a few studies exist on inherent and growth-induced LGX crystal defects[14]. Nearly all defects generate localized stresses in the crystal, and their effect on a specific device property will depend on their localization, their spatial extension and magnitude of the disturbance (stress).

Typical defects observed in langasites are striations, dislocations, inclusions, precipitates and voids, core and faceting. Striations are periodical (3D) short-range variations of the crystal composition that result in periodic lattice parameters fluctuations. Depending on the wafer orientation, striations may appear as rings, lines, or curves, with different propagation directions[6,7,14]. In LGX materials, the congruent melting composition does not correspond to the stoichiometric composition. During growth, this results in continuous radial and longitudinal shifts of the crystal composition[6,7]. In LGX, striations were found to induce SAW velocity variations/frequency shifts, which were dependent on the orientation of the striations with respect to the SAW propagation direction[6]. The effect of striations on LGX BAW devices is still unclear. Most common methods used to reveal striations are etching and X-ray synchrotron topography, both very sensitive to any crystal lattice perturbation/strain.

Inclusions are very common in Czochralski-grown LGX crystals. Their concentration may vary from a few to $< 10^6/cm^3$, depending on specific growth conditions. Their spatial distribution within the bulk of the crystals may follow the patterns of other defects, like (internal) facets, as they segregate unevenly between faceted and non-faceted crystal regions[14]. A stress field surrounds inclusions, and they are sources of dislocations. Inhomogeneous density and distribution of inclusions and dislocations lead to inhomogeneous stress fields, which will affect the acoustic properties of the material. In LGX, inclusions can be prevented by proper adjustment of growth conditions.

LGX crystals also show faceting and core structure. The core is located in the center of the crystal, and its diameter depends on the shape of the growth interface. Facets induce impurities/inclusions and dopant segregation, and generate strain especially at the boundaries where facets with different directions met. In LGX, the highest stressed regions are observed along the core region.

In this study, we report on the growth of high-quality large-diameter LGT and CTGS crystals. We demonstrate for the first time clear and transparent LGT crystals that are free of inclusions and dislocations. Such high-quality crystals allow reproducibility in bulk (BAW) and surface acoustic wave (SAW) devices, which is required for commercialization.

EXPERIMENTAL

The LGT and CTGS crystals were grown by conventional RF-heating (50kW, 10kHz, Pilar industries) Czochralski technique with automatic diameter control. Iridium crucibles of 85 x 90 and 80 x 85 (internal diameter x height, in mm) were used for the growth of 2" LGT and 1.5" CTGS, respectively. The starting oxides were 5N purity La_2O_3, Ga_2O_3, $CaCO_3$, and Al_2O_3, and 4N purity Ta_2O_5. These oxides were dried and mixed in stoichiometric ratio. This composition was slightly adjusted from one growth run to another, e.g. during the optimization process, in an attempt to approach the congruent composition and thereby reduce the defect structure of the crystal. The crystals were grown in pure nitrogen, at a pull rate between 1 and 1.5mm/h, and at a rotation rate between 10-30rpm, depending on crystal diameter. The LGT crystals were grown along the three crystal physical (orthonormal) axes X, Y, Z, which correspond to the crystallographic growth directions $[2\,\overline{1}\,\overline{1}\,0]$, $[0\,1\,\overline{1}\,0]$, $[0\,0\,.1]$, respectively, according to IEEE Standard[15]. After the growth, the LGT crystals were annealed in-situ in nitrogen atmosphere at about 1350°C, and then slowly cooled to room temperature at a rate of 50 to 80°C/h.

After the growth, the crystal boules were cut through their length into slices. These slices were polished on both sides and further inspected for potential defects. Details concerning the characterization procedure of macroscopic defects in bulk LGX and wafers have already been published[14], and will not be further discussed herein.

RESULTS

LGT crystals

2 inch LGT crystals were grown along the three main physical axes X, Y, and Z. The growth of Z-oriented boules was easier, in terms of growth process control, than growth along Y and X. Y-oriented growth was found to be the most difficult. In addition, X- and Y-grown boules had tendency to crack after the growth, during cooling to room temperature. In terms of crystal growth, the main reason to investigate different growth directions reside in the fact that facets play an important role in the overall crystal defect (distribution) structure. Such investigations allow the selection of the best growth direction in terms of minimization of defects.

From observation of the LGT crystals grown along different axes, it is clear that this material has a strong tendency to form facets. Fig. 1 shows an X-oriented LGT crystal.

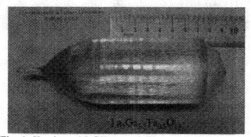

Fig. 1. X-oriented LGT crystal (X = growth direction).

Fig. 2 a) shows a Z-oriented LGT crystal, and b) the slices cut from this crystal. Z-oriented LGT crystals have a typical near-hexagonal shape, with 6 visible $\{1\,0\,\bar{1}\,0\}$ facets on the side of the cylindrical part of the boules, and X-oriented LGT crystals usually show two $\{1\,0\,\bar{1}\,0\}$ side facets, which is consistent with crystal symmetry.

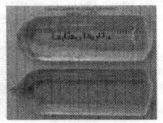

Fig. 2. a) Z-oriented LGT crystal   b) Slices cut along the growth axis.

In LGT, the basis plane $(0\,0\,0\,1)$ is a morphologically important plane that may be part of the equilibrium crystal facets. Therefore, facets are also observed on that plane in LGT. This is in contrast to quartz whose Z-planes are not part of the growth form. Z-plane facets are most apparent in the core region of Z-oriented LGT crystals. In order to investigate defects and stress distribution, polished middle slices (of about 7mm thickness) were cut along the whole crystal length and observed between crossed polars with monochromatic light,as shown in fig. 3.

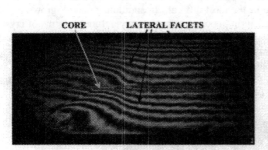

Fig. 3. Center slice of a Z-oriented LGT boule observed under crossed polars with monochromatic light (left), and microphotography of the core region with faceting (right).

The core region can easily be recognized, as well as additional $\{1\,0\,\bar{1}\,1\}$ "internal" facets. The detailed structure of the core and internal facets can be observed under a microscope[14] with side illumination. In the microphotography of fig. 3 (right), these lateral facets appear as lines. However, they have to be regarded as planes within the bulk of the crystal.

Macrosteps and facets lead to the formation of macrostep-induced (or kinetic) striations, which have a different origin than the compositional striations discussed before. They usually appear as straight lines as they propagate along a given crystallographic direction/plane. Such striations were also observed in these LGT crystals. In lithium niobate it has been found that the velocity of propagation of acoustic waves varies at regions where macrosteps/facets with different directions met. Thus, it is important to understand and prevent faceting in LGT growth.

Besides etching and optical characterization, the polished crystal slices were imaged by X-ray synchrotron topography. X-ray synchrotron topography, in its various techniques, is one of the most powerful methods for imaging and characterizing individual defects such as dislocations and striations in relatively large and perfect single-crystals. During the course of this research and optimization of the growth process, the quality of the crystals could be significantly improved. On a specific LGT crystal, the striations were found vanishing radially from the crystal center towards outside, leaving large outer crystal regions free of striations. Though it is still unclear why this occurs in that particular crystal, this result is important. Another significant result is shown in fig. 4, which is a X-ray topographic image of a slice of the crystal shown in fig. 2. It is free of dislocations and inclusions, the main defect are striations.

Fig. 4. White beam synchrotron X-ray topography of a crystal slice shown in fig 2 b).

The LGT crystals were processed into Y-cut plano-convex resonators with a radius of curvature of 26.5 cm (2 diopter contour) and a blank diameter of 14 mm (0.550 inch). The nominal 5th overtone (OT) frequency is 10MHz. The contour was selected to achieve high Q for the 5th and higher overtones by allowing adequate confinement of acoustic energy away from the mounting points and blank edge for these modes. Both the calculations and the measured data show that the fundamental mode is not sufficiently confined to permit maximum Q. The $Q{\cdot}F$ product is a convenient figure of merit used to assess the performances of a piezoelectric material[16]. The $Q{\cdot}F$ product for these devices has been measured from the Fundamental to the 9th overtone. Preliminary results are summarized in Table 1, with the maximum and median values obtained for 70 resonators built from one crystal. $Q{\cdot}F$ is reported assuming a frequency F taken in MHz.

Table I. Results for Y-cut plano-convex LGT resonators

OT	Fs [MHz]	$Q\ [10^3]$		$Q{\cdot}F\ [10^6]$	
		Max	Median	Max	Median
1	2.018	184	114	0.37	0.23
3	6.030	1493	1218	9.0	7.3
5	10.052	1662	1481	16.6	14.8
7	14.065	1613	1300	22.6	18.2
9	18.037	1140		20.5	

These resonator results compare very favorably to those for quartz, where the maximum measured $Q \cdot F$ for AT-cut and SC-cut quartz is 16 million (with F taken in MHz). The fact that the LGT results for overtones 5 through 9 are above that maximum is very encouraging.

Besides crystal quality, the design and manufacturing of the resonators can have a dramatic effect on the measured performance (Q-value) of the resonators, and this is an area where improvements are still being made. Thus, the values of table 1 should not be regarded as the highest $Q \cdot F$ product that can be obtained for LGT resonators.

CTGS crystals

The growth of CTGS crystals is still at an early stage. Initial small-diameter crystals were grown in order to establish basic process and growth conditions. Fig. 5 shows a CTGS crystal and a cube prepared from that crystal for fundamental measurements. The constant and regular diameter along the whole crystal indicates that the growth process went very smoothly.

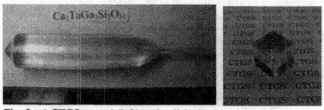

Fig. 5. a) CTGS crystal (left) and polished cube (right)

Following this promising result, CTGS crystals of larger diameter were grown. In fact, growth came out less difficult than one could have expected from the high viscosity of the melt at a melting temperature of about 1290°C. Fig. 6 shows a colorless transparent CTGS crystal of about 1.5 inch diameter. Preliminary characterization results suggest that CTGS does not show similar core/faceting problems than LGT, and etching experiments did not reveal striations. We also found that in these crystals, the (etch pit) dislocation density was extremely low (below $10^2/cm^3$). The development of CTGS crystals is still at an early stage. Further crystal characterization and device manufacture is an ongoing effort. However, from these early results, perspectives towards high-quality crystals of larger diameter for commercialization look very promising.

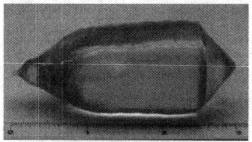

Fig. 6. Colorless transparent CTGS crystal of about 1.5" diameter.

CONCLUSIONS

High-quality colorless transparent 2" LGT and 1.5" CTGS crystals were grown by Czochralski technique and characterized. The quality of these crystals was assessed by X-ray synchrotron topography. LGT crystals free of dislocations and inclusions were obtained. In LGT, we also found striations vanishing radially from the crystal center towards outside, leaving large crystal regions virtually striations-free. The adjustment of growth conditions to prevent a specific kind of defect is a delicate balance and complex procedure, and further studies are under way to better understand the growth of these materials. The LGT crystals were processed into Y-cut plano-convex resonators, and the $Q \cdot F$ product of the devices was measured from fundamental to 9th overtone. The values obtained for the 5th, 7th and 9th overtone exceed the maximum limit of SC- and AT-cut quartz resonators. Preliminary characterization of CTGS crystal defects suggests that this material does not present the faceting/core problem of LGT. No striations were resolved by etching, and the etch pit dislocation density was found very small, below $10^2/cm^3$. Growth, characterization, and device properties of LGT and CTGS materials is ongoing. However, our early results suggest that CTGS may compete advantageously with respect to LGT, especially if CTGS crystals with larger diameter and similar high-quality can be grown.

ACKNOWLEDGMENT

The authors would like to thank M. Dudley and H.Chen at SUNYSB for imaging the slices, W.H. Horton, E. Hague and R. Helmbold from MtronPTI for the processing of the crystals and many useful discussions and advice, C. Fazi/ARL for support, and D.C. Malocha for discussions and encouragement. Financial support for has been provided by MtronPTI and through a grant from the U.S. Army.

REFERENCES

[1] R.C. Smythe, R.C. Helmbold, G.E. Hague, and K.A. Snow, "Langasite, langanite, and langatate bulk-wave Y-cut resonators," IEEE Transactions on Ultrasonics, Ferroelectrics, and Frequency Control, 47, 355- 360 (2000).

[2] R. C. Smythe, "Material and resonator properties of langasite and langatate: a progress report," in Proceedings of the 1998 IEEE International Frequency Control Symp., 761-765 (1998).

[3] Y. Kim and A. Ballato, "Force-frequency effects of Y-cut langanite and Y-cut langatate," in Proceedings of the 2002 IEEE International Frequency Control Symp.,328-332 (2002).

[4] J.A. Kosinski, R.A. Pastore, Jr., X. Yang, J. Yang, and J.A. Turner, "Stress-induced frequency shifts in langasite thickness-mode resonators," in Proceedings of the 2003 IEEE International Frequency Control Symp., 716-722 (2003).

[5] J.R. Vig, "Military applications of high accuracy frequency standards and clocks," IEEE Transactions on Ultrasonics, Ferroelectrics, and Frequency Control, 40, 522-527 (1993).

[6] R. Fachberger, T. Holzheu, E. Riha, E. Born, P. Pongratz, and H. Cerva, "Langasite and langatate nonuniform material properties correlated to the performance of SAW devices," Proceedings of the 2001 IEEE International Frequency Control Symp. 235-239 (2001).

[7] R. Fachberger, E. Riha, E. Born, W. Ruile, P. Pongratz, and S. Kronholz, "Homogeneity of langasite and langatate wafers," in Proceedings of the 2002 IEEE International Frequency Control Symp. 311-319 (2002).

[8] W.L. Johnson, S. A. Kim, D. S. Lauria, and R. C. Smythe, "Acoustic damping in langatate as a function of temperature, frequency, and mechanical contact," in Proceedings of the 2002 IEEE Ultrasonics Symposium, 961-964 (2002).

[9] W.L. Johnson, S.A. Kim, and S. Uda, "Acoustic loss in langasite and langanite," in Proceedings of the 2003 IEEE International Frequency Control Symp. p. 646-649 (2003).

[10] B.V. Mill, Y. V. Pisarevshy, and E. L. Belokoneva, "Synthesis, growth, and some properties of single crystals with the $Ca_3Ga_2Ge_4O_{14}$ structure," in Proceedings of the 1999 Joint Meeting of the EFTF and IEEE IFCS, 829-834 (1999).

[11] B.H.T. Chai, A.N.P. Bustamante, and M.C. Chou, "A new class of ordered langasite structure compounds," in Proceedings of the 2000 IEEE International Frequency Control Symp. 163-168 (2000)

[12] M. Adachi, T. Funakawa, and T. Karaki, "Growth of substituted langasite-type $Ca_3NbGa_3Si_2O_{14}$ single crystals, and their dielectric, elastic, and piezoelectric properties," in Proceedings of the 13th IEEE International Symoposium on Applications of Ferroelectrics, 411-414 (2002).

[13] Z. Wang, D. Yuan, Z. Cheng, X. Duan, H. Sun, X. Shi, X. Wei, Y. Lu, D. Xu, M. Lu, L. Pan, "Growth of a new ordered langasite structure compound $Ca_3TaGa_3Si_2O_{14}$ single crystal," J. Crystal Growth, **253**, 398-403 (2003).

[14] C. Klemenz, M. Berkowski, B. Deveaud-Pledran, and D.C. Malocha, "Defect structure of langasite-type crystals: a challenge for applications," in Proceedings of the 2002 IEEE International Frequency Control Symp., 301-306 (2002).

[15] IEEE Standard on Piezoelectricity, ANSI/IEEE Standards 176-1987, New York, 1-53 (1988).

[16] A.W. Warner, "Design and performance of Ultraprecise 2.5 mc Quartz Crystal Units", Bell Sys. Tech. J., 1193-1217 (1960).

GROWTH OF LaAlO₃ SINGLE CRYSTAL BY FLOATING ZONE METHOD AND ITS MICROWAVE PROPERTIES

Shotaro Suzuki*, Hitoshi Ohsato, Ken-ichi Kakimoto
Nagoya Institute of Technology,
Gokiso-cho, Showa-ku
Nagoya, Japan, 466-8555

Takeshi Shimada
NEOMAX Co., Ltd.
2-15-17 Egawa, Shimamoto-cho Mishima-gun
Osaka, Japan, 618-0013

Katsuhiro Sasaki, Hiroyasu Saka
Nagoya University
Furo-cho, Chikusa-ku
Nagoya, Japan, 464-0814

ABSTRACT

Lanthanum aluminate (LaAlO₃) single crystals were grown by float zone (FZ) method. (100) LaAlO₃ single crystal wafer was used as a seed. Crystal growth of LaAlO₃ without crack was achieved by the use of sintered rod with high density and homogeneity. Depending on the growth rate, the crystals showed a different color. Crystals grown in the rage of 5 to 20 mm/h exhibited a reddish-brown color, whereas crystals grown at 2 mm/h were colorless. These crystals were epitaxially grown to the seed crystal. The crystallographic information determined by IP-XRD (imaging plate X-ray diffractmeter) well agreed with the literature based on the pseudo-cubic structure. In the microstructure the grown crystals had polysynthetic twins of (100) phase on the plane perpendicular to the growth direction. The dielectric constant of each sample maintained in the vicinity of 24, while the quality factor (Qf) value depended on the growth rate. Very high Qf value of 390,000 GHz was obtained in the case of growth rate of 20 mm/h.

INTRODUCTION

LaAlO₃ single crystal has been expected as a substrate material for high-temperature superconducting microwave devices because of its low loss values and dielectric constant at high frequency. Hence, the development of LaAlO₃ single crystal with lower dielectic loss is a important subject of research.

The crystallographic data for $LaAlO_3$ has been reported as rhombohedral belonging to space group R-3m and unit cell dimensions of a = 5.357 Å and α = 60°6', which correspond to a = 3.79 Å, α = 90°5' based on a pseudo-cubic. $LaAlO_3$ has a trasition from rhombohedral to cubic symmetry above 500°C[1-5]. It is known that most of phase transitions from higher to lower symmetry structures accompany the twin formation. In the case of $LaAlO_3$, the lattice distortion involved in the cubic to rhombohedral transition results in the formation of both (100) and (110) twins. It has been reported that (100) twin is more predominant than (110) twin in $LaAlO_3$ single crystals.[2]

$LaAlO_3$ has a melting point of 2100 °C. Therefore, it is generally known that crystal growth of $LaAlO_3$ is difficult because of a limitation of crucible or furnace. Floating Zone (FZ) method is one of the effective techniques to obtain single crystal with high melting point. A cross sectional schematic of a typical crystal growth furnace is shown in Fig.1. As shown in Fig.1, the main benefit of FZ method is crucible-free technique and the high growth temperature. These points are very suitable for crystal growth of $LaAlO_3$. In this study, we use FZ method to obtain $LaAlO_3$ single crystal. The effect of growth rate on the phenomenological information and microwave dielectric properties were examined.

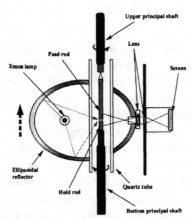

Fig.1. Schematic cross-section of the float zone furnace.

EXPERIMENTAL

La_2O_3 (Rare metallic Co,. Japan) with a purity of 99.9% and α-Al_2O_3 (Taimei chemicals Co.LTD., Japan) with a purity of 99.99% were used as starting materials for preparation of feed rods. Stoichiometric compound of the raw materials for La_2O_3 : Al_2O_3= 1 : 1 were ball-milled for

20 h. After drying, the mixture was calcined at 1450°C for 3 h. The feed rod with a circular and columnar for 8 mm diameter and 50 mm length was prepared in a silicon rubber mold using a cold isostatic pressure of approximately 200 MPa. The feed rod was sintered at 1650 °C for 4 h in air.

The growth apparatus, FZ furnace, was an infrared convergence-type image furnace with single ellipsoid mirror (NEC.Co, Japan: SC-30XS-MP). Xe arc lamp of 2 kW was set at the focal point of ellipsoid as infrared source. $LaAlO_3$ single crytal wafer (Earth chemical Co., Japan) with (100) based on pseudo-cubic unit cell was used as a seed. The sintered feed rod was suspended from the upper shaft and the seed crystal was fixed to the lower shaft. The rotation rate was 40 rpm for both the feed rod and the seed crystal. The growth rate was changed in the range from 2 to 20 mm/h. The growth atmosphere was air.

The crystalline phases were identified by X-ray diffraction (XRD; Philips, Nederland: X'Rert MPD). The crystallographic information of $LaAlO_3$ crystal was determined by IP (imaging plate) XRD (Rigaku Co., Japan: R-AXIS RAPID). The domain structure within the crystal was observed by polarizing microscope. The grown crystal rods were formed to pellet shape with diameter of 5 mm and height of 2.5 mm to measure the microwave dielectric properties. The dielectric constant (ε_r) and unloaded Q values were measured using a pair of parallel conducting Cu plates under TE_{011} mode using Hakki and Coleman's method.[6]

RESULTS AND DISCUSSION

$LaAlO_3$ crystals grown by FZ furnace

The obtained Photographs of $LaAlO_3$ crystals grown by FZ method at a grown rate of 20 and 2mm/h are shown in Fig.2 (a) and (b), respectively. In order to obtain crystal without cracks and defects, the using sintered rod should have a high density and homogeneity. In this study, we could successfully obtain a relatively crystal without cracks and defects. At growth rate in the range from 5 to 20 mm/h, the crystal exhibits a reddish brown color. Since the crystal growth was performed in air, oxygen deficiency may be the reason of a reddish brown color, as reported in the literature[3,7]. On the other hand, the growth rate in 2 mm/h caused colorless crystals. It is considered that slower growth rate maintained the stoichiometry during the crystallization. Thus, the color of crystal is closed to colorless which is original color of crystal without oxygen deficiency.

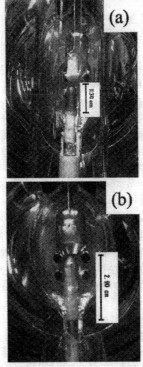

Fig.2. Photographs of LaAlO₃ crystals grown by FZ at a growth rate
of (a) 20 mm/h, and (b) 2 mm/h.

A XRD pattern for the crystal plane of perpendicular to growing direction is shown in
Fig.3. The resulting pattern showed only those reflections belonging to the {100} based on the
pseudo-cubic cell indexing of LaAlO₃. Therefore, it is clear that the crystals were epitaxially
grown to the seed crystal. An IP X-ray diffractometer determined that the lattice constants were a
= 3.795(7) Å, α = 90.11(9) °, crystal system was trigonal and space group was R-3m. This lattice
constants are based on a pseudo-cubic. These crystallographic parameters well agreed with the
reported data.[1-3]

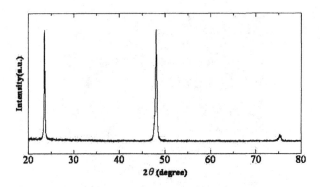

Fig.3. XRD pattern of LaAlO₃ crystal on the plane perpendicular to [100] of pseudo-cubic cell. The growth rate is 20 mm/h.

Twin structure

Figure 4(a) shows a optical photograph of LaAlO₃ crystal wafer observed by polarization microscope in crossed polar. The wafer was prepared by cutting the crystal at plane perpendicular to growth direction. The observation revealed the existence of polysynthetic twins on the plane of wafer. When observing stage of the microscope was turned 45°, these twins were extincted same time (Fig.4(b)). The retardation in this plane was also shown in Fig.4.(c). Since addition and subtraction are next to each other, optic axes of polysynthetic twins are orthogonal. Kim and Berkstresser reported that pseudo-cubic LaAlO₃ has (100) or (110) twins [2,3]. As revealed in Fig.3, growth plane has only {100} component of the LaAlO₃. Thus, these polysynthetic twins is considered as (100) twins.

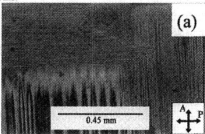

Fig.4. Optical photographs of the crystals observed by a polarizing microscope. The views of planes are perpendicular to [100]
(a) The diagonal position (b) The extinction position (c) Using a test plate.

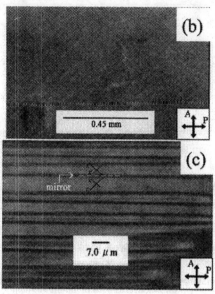

Continued

Microwave dielectric properties

The microwave dielectric properties of crystals were shown in Fig.5 as a function of growth rate. The dielectric constant of each sample exhibited in the vicinity of 24. On the other hand, the quality factor (Qf) value varied with growth rate. The highest Qf value of 390,000GHz was obtained in the case of growth rate of 20 mm/h. In contrast, the slowest growth rate of 2 mm/h, colorless crystals brought about the lowest Qf value of 140,000GHz. Although it was expected that colorless crystal achieved by the slow growth rate results in high Qf value because of decrease of oxygen vacancy in the crystal, opposite result was obtained in this study. The reason is not clear, yet. However, as shown in Fig.4, grown crystal had a twin structure. Therefore, a difference of inhomogeneity such as twin structure may affect to the dielectric properties. In the near future, further work will clarify this reason.

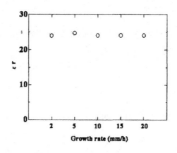

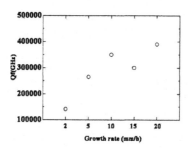

Fig.5. The microwave dielectric properties of LaAlO₃ crystals as a function of growth rate.

CONCLUSION

LaAlO₃ single crystal rods were grown by FZ method. The grown crystals displayed a reddish brown color when the growth rate was in the range from 5 to 20 mm/h. On the other hand, the slowest growth rate of 2 mm/h resulted in colorless crystal. A XRD pattern on the plane perpemdicular to growth direction revealed that obtained single crystals were grown to <100> direction with the same orientation of seed crystal wafer. LaAlO₃ crystals had polysynthetic twins of (100) phase perpendicular to growth direction. In spite of evolution of twins, the crystals exhibited very high Qf value compared with ordinary LaAlO₃ ceramics. The highest Qf value of 390,000GHz was obtained in the case of growth rate of 20 mm/h.

ACKNOWLEDGEMENTS

This work was supported by a grant from the NITECH 21st Century COE Program "World Ceramics Center for Environmental".

183

REFERENCES

[1] S.Geller and V.B.Bala, "Crystallographic Studies of Perovskite-like Compounds. II. Rare Earth Aluminates," *Acta crystallographica*, **9**, 1019-25 (1956).

[2] C.H.Kim, J.W.Jang, "Ferroelastic twins in LaAlO$_3$ polycrystals" *Physica B*, **262**, 438-43 (1999).

[3] G.W.Berkstresser and A.J.Valentino, "Growth of single crystals of lanthanum aluminate" *Joulnal of Crystal Growth.*, **109**, 467 -71 (1991)

[4] T.A.Vanderah, C.K.Lowe-Ma and D.R.Gagnon, "Synthesis and Dielectric Properties of substituted Lanthanum Aluminate," *Journal of the American Ceramic Society*, **77**[12], 3125-30 (1994)

[5] H.Fay and C.D.Brandle, in: *Crystal Growth* (Pergamon, Oxford, 1967) p.51

[6] B.W.Hakki and P.D.Coleman: *IEEE transactions on microwave theory and techniques*, **MTT-8**, 402-10 (1960)

[7] I.H.Jung and C.S.Lim, "GROWTH AND CHARACTERIZATION OF LaAlO$_3$ SINGLE CRYSTALS BY THE TRAVELING SOLVENT FLOATING ZONE METHOD" *Ceramic Transactions*, **100**, 154-64 (1999)

General Topics in
Electronic Ceramics

EFFECTS OF NIOBIUM ADDITION ON MICROSTRUCTURAL AND ELECTRICAL PROPERTIES OF LEAD ZIRCONATE TITANATE SOLID SOLUTION (PZT 95/5)

Pin Yang, James A. Voigt, Mark A. Rodriguez, Roger H. Moore, and George R. Burns
Sandia National Laboratories
Albuquerque, New Mexico 87185-0959

ABSTRACT

The impacts of small niobium additions to processing, microstructure, and electrical properties in the Zr-rich lead zirconate titanate ceramics (PZT 95/5) were investigated. The influence of niobium content on dielectric responses and the characteristics of ferroelectric behaviors, as well as the relative phase stability and the hydrostatic pressure induced ferroelectric-to-antiferroelectric phase transformation are reported. Results indicate that increasing the niobium concentration in the solid solutions enhances densification, refines the microstructure, decreases dielectric constant and spontaneous polarization, and stabilizes the ferroelectric phase. The stabilization of ferroelectric phase with respect to the antiferroelectric phase near PZT 95/5 composition dramatically increases the pressure required for the ferroelectric-to-antiferroelectric phase transformation. These observations were correlated to the creation of A-site vacancies and a slight modification of the crystal structure. The importance of these composition-property relationships on device application will be presented.

INTRODUCTION

The perovskites $PbZrO_3$ and $PbTiO_3$ form a continuous series of solid solutions over the whole composition range.[1] These solid solutions, commonly referred as PZTs, are rich in variety of ferroelectric (FE) and nonferroelectric phase transformations that can be induced by changes in composition, temperature, electric field, or pressure (stress). These materials also exhibit large spontaneous polarization and piezoelectric coefficients. These factors, along with the availability in high-quality ceramic form, are responsible for their widespread technological applications.[1]

The specific composition of interest is near $Pb(Zr_{0.95}Ti_{0.05})O_3$ (or PZT95/5) where the free energy difference between ferroelectric and antiferroelectric (AFE) phases is very small and the phase stability can be altered in the presence of an external electric field[2] or pressure.[3] This study focuses on the processing-microstructure-property relationships as a small amount of niobium (Nb) is introduced into PZT 95/5 ceramics. Special emphasis is placed on the effect of Nb addition on the hydrostatic pressure induced FE-to-AFE phase transformation for power supply applications.

The introduction of B site donor cations, such as Nb^{5+} or Sb^{5+}, into PZT ceramics creates A-site vacancies in the perovskite lattice and leads many desirable piezoelectric properties and improved ferroelectric characteristics for compositions near the morphotropic phase boundary.[1,4,5] For power supply applications, the ferroelectric properties and the relative phase stability between FE and AFE phases are more important. In this study, we focused on these issues for Nb modified PZT 95/5 ceramics and compare the changes of ferroelectric properties to those reported for ceramics near the morphotropic phase boundary.

EXPERIMENTAL PROCEDURE

The niobium modified PZT 95/5 powders were prepared by precipitating a lead acetate and Zr, Ti, Nb n-butoxide-glacial acetic acid solution with an oxalic acid-n-propanol solution.[6] The specific compositions used for this study were $Pb_{0.9935}(Zr_{0.953}Ti_{0.047})_{0.987}Nb_{0.013}O_3$, $Pb_{0.9910}(Zr_{0.953}Ti_{0.047})_{0.982}Nb_{0.018}O_3$ and $Pb_{0.9885}(Zr_{0.953}Ti_{0.047})_{0.977}Nb_{0.023}O_3$, where the molar fraction of Nb concentration was varied from 0.013 to 0.023. Upon cooling, ceramics of these compositions transform from a high temperature paraelectric phase of the ideal cubic pervoskite structure to a high temperature rhombohedral FE phase (FE_{RH}; space group R3m), and finally to a low temperature rhombohedral FE phase (FE_{RL}; R3c). Under a hydrostatic pressure, the material can be induced into an AFE_o orthorhombic (Pbam) structure.[3] An additional 0.5 mole % lead was added to the formulation to compensate for lead loss during the process. These chemically prepared powders were first pyrolyzed at 400°C and then calcined at 900°C for 16 hours. An appropriate amount of pore former (15 μm Lucite; spherical, polymerized methyl methacrylate) was added and mixed with the dry calcined PZT powder, then a 2 wt% binder solution (HA4, acrylic binder (Allied Colloid)) was sprayed upon the tumbling powder. Approximately 70 grams of the granulated powder was then fed into a tool steel die and uniaxially pressed at 96 MPa. Ceramic compacts were slowly heated up to 750°C and held for 4 hours to burn out all the organic additives. To prevent lead loss during the high temperature sintering, a double crucible technique with 95/5 PZT powder was used as an atmospheric control. Sintering was performed at 1350°C for 6 hours. Typical weight loss was controlled to less than 0.2%. The Archimedes method was used to measure the densities of the specimens. These sintered billets were then ground, sliced and electroded with a thick film silver metallization (DuPont 7095).

The phases and crystalline structure of the calcined powders were examined using X-ray diffractometry (XRD, Siemens, D500) with a 2θ angle range from 20° to 100°. A step scan with a step size of 0.04° was used with a counting time of 20 s/step. Structural refinement was performed with TOPAS software (Bruker, Madison, WI). Although the crystalline structure of both FE_{RH} and FE_{RL} phases can be presented on a rhombohedral basis, the lattice parameters of the refined data are discussed in terms of a hexagonal setting for convenience. The composition of these powders was determined by an inductively coupled plasma (ICP), atomic emission spectroscopy technique, using a reference powder batch to monitor the variations from batch to batch. Chemical analysis results show that the concentration of all cations in these solid solutions was tightly controlled (not shown). The microstructures of the chemically etched surface were studied using a scanning electron microscope (SEM, Hitachi S-4500) with a backscattered electron (BSE) detector to identify possible existence of a second phase. The temperature dependence of the dielectric constant was measured at temperatures ranging from − 20° to 115°C with a heating and cooling rate of 3°C/min at 10 kHz, using an impedance analyzer (HP4192A, Hewlet-Packard, Palo Alto, CA). Dielectric hysteresis measurements were made with a modified Sawyer-Tower circuit at room temperature. The pressure induced FE-to-AFE phase transformation was measured at room temperature. Hydrostatic pressure was increased at a rate of 10.3 MPa/s and the amount of charge release was monitored by a capacitor in series with the PZT ceramic.

RESULTS AND DISCUSSION
Structure Modification
Fig.1 gives the Rietveld refinement fitting for a typical 0.018 Nb modified PZT 95/5 powder, showing calculated (+), observed, and difference pattern (bottom). The fitting looks

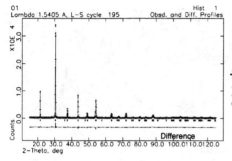

Fig. 1 Rietveld refinement fitting of Nb = 0.018 powder showing calculated (+), observed and difference patterns.

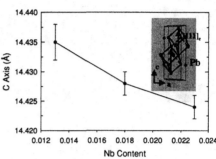

Fig. 2 Effect of Nb concentration on the c axis of a hexagonal unit cell.

quite good and the residual R_p values for all the compositions in this study (Nb = 0.013, 0.018, and 0.023) were typically in the 8% range, indicating a good match of observed to calculated patterns. The only clear trend observed in this analysis is a systematic contraction in the c axis (in a hexagonal setting) with increasing of Nb concentration (see Fig. 2). The substitution of a high valence, intermediate size Nb ion ($Nb^{5+}[6]$ = 0.78 Å) with a mix of smaller ($Ti^{4+}[6]$ = 0.745 Å) and larger ($Zr^{4+}[6]$ = 0.86 Å) cations in the B-sites would lead an increase in lattice parameters; however, the effect is compensated by the creation of large A-site vacancies ($Pb^{2+}[12]$ = 1.63 Å) in order to maintain local electroneutrality. Since the amount of Nb substitution is very small (from 0.013 to 0.023) in this study, the subtle change in lattice structure due to the B-site substitution and the creation of A site vacancies is difficult to discern in a primitive cell with high confidence. Among all the axes, the hexagonal c axis (polar axis which is in the body diagonal [111] direction, see insert in Fig. 2) is the most sensitive to a small distortion in the lattice due to a geometrical magnification. For example, in a cubic lattice a small change in lattice spacing in the [001] direction will be amplified ($x \sqrt{2}$) in the face diagonal direction [001], and will be enlarged more ($x \sqrt{3}$) in the body diagonal direction [111]. Converting from a primitive rhombohedral basis to a hexagonal setting will add another magnification factor since each hexagonal unit cell consists of 6 primitive rhombohedral cells. Therefore, it is expected that a slight change in the lattice due to the creation of Pb vacancies by Nb substitution will be manifested by the contraction in the hexagonal c axis. The moderate contraction of ~ 0.010 Å as the Nb content is varied from 0.018 to 0.023 mol% can, therefore, be attributed to the creation of Pb vacancies by Nb substitution in the PZT 95/5 solid solution.

Microstructure and Densification

The addition of Nb to the PZT solid solution below the solubility limit (~ 7 at.%) has been known from many previous studies[8] to lead to a refined microstructure and a higher sintered density for bulk ceramics. Figure 3 shows the changes of microstructure of PZT ceramics with increasing Nb content. The Nb concentration (mol%) varied from (a) 0.013 to (b) 0.018, and (c) 0.023 in this study. The spherical pores (~15 μm) due to the pore former additive are clearly seen in these photomicrographs. Samples were slightly etched to reveal the grain and domain structure. It is obvious that the grain size is strongly influenced by the additive content: increasing Nb concentration causes a decrease in grain size. The change of average grain size as

189

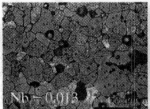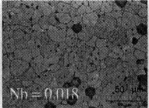

Fig. 3 The SEM images of PZT 95/5 ceramics with different Nb content.

a function of Nb content is given in Fig. 4. The grain size was determined by a linear intercept method from Fig. 3. The average grain sizes for compositions with 0.013 Nb and 0.023 Nb are 14.40 µm and 6.92 µm, respectively. This large change of grain size as a result of a small Nb modification illustrates the dramatic effect of Nb on the microstructure refinement. This observation is consistent with the strong grain size reduction observed in the PZT morphotropic phase boundary due to the substitution of A-[7] or B-site[8] cations. It should also be noted here that the general microstructure refinement effect observed in bulk ceramics can not applied to the sol-gel derived thin layers[9] where the Nb addition stabilizes the transient pyrochlore phase during the in-situ reaction and promotes lateral grain growth and columnarity.

Fig. 5 shows the effect of Nb on the bulk density of PZT 95/5 ceramics. Data indicate that the bulk density increases with increasing Nb content. In addition, the number of intra-granular pores decreases (see Fig. 3). In the highest Nb content sample (Fig. 3 (c) Nb = 0.023), almost all of the intragranular pores were eliminated except one or two smaller pores observed within a few larger grains. Earlier work[10] has shown that Nb modified PZT has always achieved a much higher density than the pure PZT ceramics at different heating rates. Comparing the microstructures shown in Fig. 3, the increase in bulk density due to the Nb addition can be attributed to the removal of intragranular pores in the sintered ceramics. The general behavior observed, including the enhanced densification, refining of the microstructure, and the removal of intragranular pores by the Nb modification, can be best described by the reduction of grain boundary mobility, similar to that observed in the Y_2O_3 system[11] where minor ThO_2 was added as a grain growth inhibitor. Since grain boundary mobility strongly depends on numerous factors, such as local grain boundary structure, porosity size,[12] and composition, as well as the

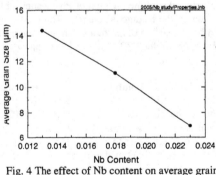

Fig. 4 The effect of Nb content on average grain size.

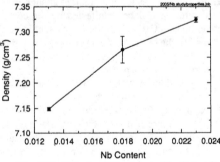

Fig. 5 The effect of Nb content of bulk density of sintered ceramics.

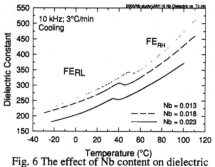

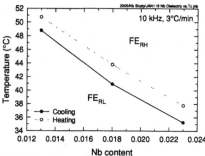

Fig. 6 The effect of Nb content on dielectric constant of PZT 95/5 ceramics.

Fig. 7 The effect of Nb on the relative stability of FE_{RH} and FE_{RL} phases.

electrostatic charge, temperature and impurity solutes, the fundamental mechanism governing the observed behavior is beyond the scope of this work.

Dielectric and Ferroelectric Properties

Fig. 6 shows the effect of Nb addition on the dielectric constant and the relative thermal stability between FE_{RH} and FE_{RL} phases. Data were collected at 10 kHz during the cooling cycle with a cooling rate of 3°C per minute. Results indicate that increasing the Nb content in PZT95/5 ceramic lowers the dielectric constant and stabilizes the FE_{RH} phase to a lower temperature. In addition, a thermal hysteresis of ~3°C was observed during the structure phase transformation between the FE_{RH} and FE_{RL} upon the heating and the cooling cycles (see Fig. 7). Note within this small composition modification, there is a significant downwards shift in the phase transformation temperature (~ 14°C) as the Nb content changes from 0.013 to 0.023. This immediately suggests that adding Nb into the Zr-rich PZT solid solutions has a great impact on the relative phase stability between two rhombohedral phases and changes the phase boundary in the PZT phase diagram, as has been demonstrated by pulsed neutron diffraction.[13]

The effect of small compositional modifications on the dielectric constant is a complex issue in ferroelectrics. The major premise on the altering of dielectric constant (or dielectric susceptibility dP/dE) relies on the change of the overall polarizability of the material due to small compositional modifications in the lattice. The decrease of highly polarizable Pb^{2+} cations along the polar axis due to substitution of Nb seems to be a sensible explanation for the moderate repression in dielectric constant in Fig. 6. Another interpretation can result from a different premise. Polarization in the rhombohedral ferroelectric perovskite lattice depends on the ionic displacement of B-site and A-site cations along the polar axis with respect to a fixed oxygen plane (O at 18b (x,y,z = 1/12)). The substitution of a medium size Nb^{5+} cations in the B-site together with the creation of A-site vacancies can effectively reduce the dynamic strain[14] permitted for the cations along the polar axis under a weak dc excitation condition (such as a dielectric constant measurement), and leads to a lower dielectric susceptibility. A previous work on the effect of lead stoichiometry on the dielectric response of PZT ceramics with the same composition[15] has found that the dielectric constant increases with Pb content. These results suggest that the dielectric constant seems to be quite sensitive to the Pb content and the dynamic strain of cations along the polar axis for ceramics near PZT 95/5 composition.

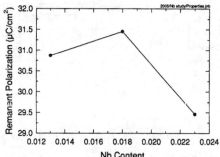

Fig. 8 The change of remanent polarization as a function of Nb concentration.

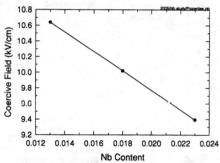

Fig. 9 The effect of Nb on the coercive field.

The structural phase transformation between FE_{RH} (R3m) and FE_{RL} (R3c) phases can be characterized in terms of antiphase rotation and tilting of oxygen octaheda, as well as the oxygen anions coupling to cation shifts along the [111] axis.[16, 17, 18] Yang et. al. studied the relative phase stability between these two ferroelectric phases under a strong dc bias condition[19] and observed that the high temperature to low temperature transformation was trigged by a short-range electrostatic interaction between the lead cation and oxygen anions at the adjacent cell along the polar axis followed by the oxygen octahedral rotation/tilting. The creation of Pb vacancies due to the Nb substitution in lattice can effectively reduce this short-range electrostatic interaction. This structure modification could contribute to the down shift of FE_{RH}-to-FE_{RL} transformation temperature as observed in Fig 6.

Dielectric hysteresis (P-E) data were collected at 0.05 Hz at room temperature. Results of the remanent polarization and the coercive field from ferroelectric polarization reversal for PZT95/5 ceramics with various Nb content are given in Fig. 8 and Fig. 9, respectively. The data indicate that as the amount of Nb increases from 0.013 to 0.018 the remanent polarization first increases slightly than drops to a lower value. However, the coercive field decreases monotonically. The decrease in coercive field and an increase in the squareness of dielectric hysteresis loop are attributed to the increase in domain wall mobility due to the creation of A-site vacancies in the lattice. Ceramics with these characteristics are generally referred as the "soft" PZTs in the literature. However, in essence, the coercive field should be directly related to the magnitude of the dielectric dipole moment in the ferroelectrics, where a greater electric field is required to rotate a larger dipole in the field direction. This argument is clearly demonstrated by the increase in coercive field as a ferroelectric material is cooling down from the Curie temperature (an example can be found in ref. 19). Therefore, the higher coercive field observed for the lowest Nb content sample (0.013 Nb) suggests that the material should have the highest remanent polarization, which is partially inconsistent with data in Fig. 8. Attempts to resolve this conflict by estimating the spontaneous polarization through the ionic displacements in the lattice by X-ray structure refinement data, were unfruitful (in the range of 33 to 35 $\mu C/cm^2$ range without a clear trend) due to the uncertainty of the A-site cation position[16, 20] and subtle changes in atomic displacements. It is believed that this slightly lower remanent

192

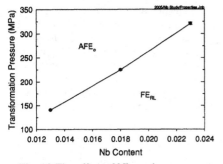

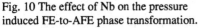

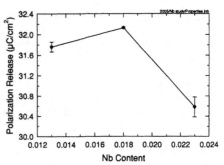

Fig. 10 The effect of Nb on the pressure induced FE-to-AFE phase transformation.

Fig. 11 The amount of polarization release as a function of Nb content.

polarization for Nb=0.013 (in comparison with Nb = 0.018) could be attributed to a relatively low bulk density as well as the inability to pole all the polar domains in the field direction in the presence of a large amount of intragranular porosity in the microstructure.

Pressure-Induced FE-to-AFE Phase Transformation

A thermal mechanical measurement on PZT 95/5 ceramics without Nb shows a thermally induced FE_{RL}-to-AFE_o phase transformation at -40 °C, with a large negative transformational strain[21] (i.e., -0.23% linear shrinkage in the polar axis). A Raman study on 1 wt% Nb modified PZT97.5/2.5 indicates that the temperature required to induce the AFE_o phase is at -170°C.[22] For the compositions (Nb $\geq$ 0.018) studied in this paper, the thermally induced transformation cannot be observed even when the ceramics were cooled down to − 261°C.[23] These evidences indicate that a small amount Nb can effectively stabilize the low temperature rhombohedral FE phase and suppress the thermally induced FE_{RL}-to-AFE_o phase transformation.

Fig. 10 illustrates the effect of Nb on the transformation pressure for a hydrostatic pressure induced FE_{RL}-to-AFE phase transformation at room temperature. A previous study[24] has shown that dielectric properties and hydrostatic pressure induced transformation will not be affected if the average grain size of ceramic is greater than 2 µm. Since the average grain sizes in this study are much greater than this value, the effect of grain size on transformation pressure becomes trivial. Results show that the transformation pressure increases monotonically with Nb content. Note a small change in Nb from 0.013 to 0.023 can drastically increase the transformation pressure from 141 MPa to 321 MPa. The increase of pressure required to induce the FE_{RL}-to-AFE_o transformation indicates that the free energy difference between these two phases becomes greater as Nb is added to the solid solution. As a result, a greater external work (by the pressure) is required to induce the FE_{RL}-to-AFE_o phase transformation.

Fig. 11shows the amount of polarization release associated with the hydrostatic pressure induced transformation. This result mirrors the amount of remanent polarization stored on the electrodes during the poling process as shown in Figure 8.

CONCLUSIONS

The effects of small Nb modifications in PZT 95/5 ceramics were studied. Most observations are consistent with results reported for compositions near the PZT morphotropic phase boundary. The changes induced by the Nb additions are correlated to the creation of A-site vacancies and the crystal structure modification. The creation of A-site vacancies enhances bulk diffusion, promotes densification, and reduces the amount of intragranular pores. In addition, Nb additions refine the microstructure. The contraction of the polar axis due to the creation of A-site vacancies represses the A-site and B-site cation displacements and renders a lower remanent polarization and a smaller coercive field. Nb additions also tend to decrease the FE_{RH}-FE_{RL} transformation temperature, suppress the thermally induced FE_{RL}-to-AFE_O phase transformation, and dramatically increase the FE_{RL}-to-AFE_O phase transformation pressure. Therefore, the control of Nb concentration in the PZT 95/5 ceramics is important to insure consistent device performance.

ACKNOWLEDGEMENT

The authors would like to acknowledge Diana L. Moore and Mike Hutchinson for powder synthesis and ceramic fabrication, as well as Chad S. Watson for microstructure evaluation. Sandia is a multiprogram laboratory operated by Sandia Corporation, a Lockheed Martin Company, for the United States Department of Energy under Contract No. DE-AC04-94AL85000.

REFERENCES

[1] B. Jaffe, W. R. Cook, Jr., and H. Jaffe, *Piezoelectric Ceramics*, Academic Press, New York (1971), Chapter 7.

[2] P. Yang and D. A. Payne, "Thermal Stability of Field-Forced and Field-Assisted Antiferroelectric-Ferroelectric Phase Transformations in Pb(Zr,Sn,Ti)O$_3$," *J. Appl. Phys.*, **71** 1361-7 (1992).

[3] I. J. Fritz and J. D. Keck, "Pressure-Temperature Phase Diagrams for Several Modified Lead Zirconate Ceramics," *J. Phys. Chem. Solids*, **39**, 1163-7 (1978).

[4] D. Berlincourt, "Piezoelectric Ceramic Compositional Development," *J. Acoust. Soc. Am.*, **91** 3034-40 (1992).

[5] R. Gerson, "Variation in Ferroelectric Characteristics of Lead Zirconate Titanate Ceramics Due to Minor Chemical Modifications," *J. Appl. Phys.*, **31**, 188-94 (1960).

[6] J.A. Voigt, D.L. Sipola, B.A. Tuttle, and M. Anderson, "Non-aqueous Solution Synthesis Process for Preparing Oxide Powders of Lead Zirconate Titanate and Related Materials," U. S. Pat. No. 5 908 802, June 1, 1999.

[7] M. Hammer and M. J. Hoffmann, "Sintering Model for Mixed-Oxide-Derived Lead Zirconate Titanate Ceramics," *J. Am. Ceram., Soc.*, **81**, 3277-84 (1998).

[8] M. Pereira, A. G. Peixoto, and M. J. M. Gomes, "Effect of Nb Doping on the Microstructural and Electrical Properties of PZT Ceramics," *J. Europ. Ceram. Soc.*, **21**, 1353-1356 (2001).

[9] R. D. Klissurska, K. G. Brooks, I. M. Reaney, C. Pawlaczyk, M. Kosec, and N. Setter, "Effect of Nb Doping on the Microstructure of Sol-Gel-Derived PZT Thin Films," *J. Am. Ceram. Soc.*, **78**, 1513-20 (1995).

[10] J. Ryu, J-J Choi, and H-E Kim, "Effect of Heating Rate on the Sintering Behavior and Piezoelectric Properties of Lead Zirconate Titanate Ceramics," *J. Am. Ceram. Soc.*, **84**, 902-904 (2001).

[11] W. D. Kingery, H. K. Bowen, and D. R. Uhlmann, *Introduction to Ceramics*, Chapter 10, John Wiley & Sons, New York, 1976.

[12] V. Tikare, M. A. Miodownik, and E. A. Holm, "Three-Dimensional Simulation of Grain Growth in the Presence of Mobile Pores," *J. Am. Ceram. Soc.*, **84**, 1379-85 (2001).

[13] B. Noheda, J. A. Gonzalo, and M. Hagen, "Pulsed Neutron Diffraction Study of Zr-Rich PZT," *J. Phys.: Condens. Matter*, **11**, 3959-65 (1999).

[14] J. N. Lin and T. B. Wu, "Effects of Isovalent Substitutions on Lattice Softening and Transition Character of $BaTiO_3$ Solid Solutions," *J. Appl. Phys.*, **68**, 985-93 (1990).

[15] P. Yang, J. A. Voigt, S. J. Lockwood, M. A. Rodriguez, G. R. Burns, and C. S. Watson, "The Effect of Lead Stoichiometry on the Dielectric Performance of Niobium Modified PZT 95/5 Ceramics," Ceramic Transitions, **105**, 289-297 (2004).

[16] D. L. Corker, A. M. Glazer, R. W. Whatmore, A. Stallard, and F. Futh, "A Neutron Diffraction Investigation into the Rhombohedral Phases of the Pervoskite Series $PbZr_{1-x}Ti_xO_3$," *J. Phys. Condens. Matter.*, **10**, 6251-6269 (1998).

[17] A. M. Glazer, S. A. Mabud, and R. Clark, "Powder Profile Refinement of Lead Zirconate at Several Temperatures I. $PbZr_{0.9}Ti_{0.1}O_3$," *Acta Cryst.*, **B34**, 1060-5 (1978).

[18] Z. Jirak and T. Kala, "Crystal Structures of Ferroelectric Phases $F_{R(LT)}/F_{R(HT)}$ in $PbZr_{0.75}Ti_{0.25}O_3$ Solid Solutions and Their Dependence on Temperature," *Ferroelectrics*, **82**, 79-84 (1988).

[19] P. Yang, M. A. Rodriguez, G. R. Burns, M. E. Stavig, and R. H. Moore, "Electric Field Effect on the Rhombohedral-Rhombohedral Phase Transformation in Tin Modified Lead Zirconate Titanate Ceramics," *J. Appl. Phys.*, **95** 3626-32 (2004).

[20] S. Teslic, T. Egami and D. Viehland, "Local Atomic Structure of PZT and PLZT Studied by Pulsed Neutron Scattering," **57**, 1537-43 (1996).

[21] P. Yang, unpublished data (2004).

[22] S. Kojima, N. Ohta, and X. Dong, "Stability of Antiferroelectric and Ferroelectric Phases in Low Ti Concentration $PbZr_{1-x}Ti_xO_3$," *Jpn. J. Appl. Phys.*, **38**, 5674-8 (1999).

[23] M. Avdeev, J. D. Jorgensen, S. Short, B. Morosin, E. L. Venturini, P. Yang, and G. A. Samara, "Pressure-Induced Ferroelectric to Antiferroelectric Phase Transition of PZT95/5(2Nb): A Neutron Powder Diffraction and Dielectric Study," to be submitted to *Phys. Rev. B.* (2005).

[24] B. A. Tuttle, J. A. Voigt, T. W. Scofield, P. Yang, D. H. Zeuch, and M. A. Rodriguez, "Dielectric Properties and Depoling Characteristics of $Pb(Zr_{0.95}Ti_{0.05})O_3$ Based Ceramics: Near-Critical Grain Size Behavior,"; 55-58 in Proceedings of the 9th U.S. and Japan Seminar on Dielectric and Piezoelectric Ceramics (Okinawa, Japan, November 2-5, 1999).

ENHANCED DENSITY AND PIEZOELECTRIC ANISOTROPY IN HIGH TC PbNb$_2$O$_6$ BASED FERROELECTRIC CERAMICS

Ducinei Garcia, Michel Venet, Aline Vendramini, José Antonio Eiras
Federal University of São Carlos
Rod. Washington Luis, km 235
São Carlos, SP, 13565-905

Fidel Guerrero
University of Oriente
Patricio Lumumba. C.P. 90500.
Santiago de Cuba. Cuba

ABSTRACT

In this work, conventional synthesis was employed to obtain highly dense Ti or Ba-doped PbNb$_2$O$_6$ ceramics. The ferroelectric phase (at room temperature) was obtained through a heating treatment at 1300°C during 1h, after calcination process. The compensation of the lead oxide loss was made by the ideal addition of PbO excess prior calcination and sintering steps. The sintering parameters as time and temperature were studied to optimize the firing conditions. Ceramics of orthorhombic structure and high density were obtained. The temperature dependence of the dielectric properties and the piezoelectric coefficients for the ceramics were analyzed. High Curie temperatures and high piezoelectric anisotropy were found, especially for the lower content Ti doped PN ceramics.

INTRODUCTION

PbNb$_2$O$_6$ (PN) is a ferroelectric material with a tungsten-bronze structure and orthorhombic symmetry at the ferroelectric phase.[1-3] Due to its high Curie temperature (Tc) and low quality factor (Q), this material has potentiality to be used in the fabrication of ultrasonic transducers, for high temperature applications, such as flow detectors devices (where the Pb(Zr,Ti)O$_3$ and other piezoelectric materials can not be used).[4]

The fabrication of PN ceramics is extremely difficult since appropriated heat treatments are necessary in order to obtain the orthorhombic ferroelectric phase.[3,5,6] Also, cracks, abnormal grain growth, and formation of undesirable phases (promoted by the PbO losses) may affect the densification of the PN ceramics, difficulting the polarization[5,6] and, consequently, their use in ultrasonic transducers.

Some elements may be used as additives for the PN synthesis. The addition of barium[7-9] (Pb$_x$Ba$_{1-x}$Nb$_2$O$_6$ or PBNx) contributes for the densification, but strongly reduces the Curie temperature, harming the high temperature applications. Other elements, such as Pb, Ca, Mn La and Ti,[3,4,10] have been used to increase the densification. The addition of Ti not only increases the densification of the PN ceramic, but also increases Tc,[3] which is a desirable property for high temperature transducers. However, to obtain a single phase compound with orthorhombic ferroelectric phase, some works have reported the use of quenching procedure.[5,6] That may be inconvenient for high quantities processing.

In this work, highly densified Ti- and Ba-doped PN ceramics were prepared. The main piezoelectric properties were measured and analyzed aiming high temperature applications.

EXPERIMENTAL

Ceramic powders with nominal formula $Pb_{1+x}Ti_xNb_{2-x}O_6$ (PTN100x), with x=0.0, 0.05 and 0.10 and $Pb_{0.75}Ba_{0.25}Nb_2O_6$ (PBN75) were prepared by the solid-state reaction method. Analytical graded precursors (with an excess of 2 wt% PbO, in the case of PN and PTN), were mixed in a ball mill, containing isopropyl alcohol and stabilized ZrO_2 cylinders, during 20 h. The mixture was dried, and calcined at 1050 °C for 3,5 h and milled again for 20 h. The powders of PN and PTN were heat treated at 1300 °C for 1 h in order to obtain the room temperature orthorhombic ferroelectric phase. After the thermal treatment, 3 wt% of PbO was added to compensate lead losses during the sintering process, which occurred at 1270 °C / 4,5 h. Same sintering conditions were used for the PBN75 sample, although the addition of PbO excess and thermal treatment were not necessary to obtain the ferroelectric phase. This sample was prepared mainly to compare the effect of both dopants on PN plain composition. X-ray diffraction technique was performed to a qualitative phase analysis and to determinate the lattice parameters at room temperature, using a Rigaku diffractometer and CuK_α radiation at room temperature. Computer assisted dielectric characterization was realized as a function of the temperature, using an HP 4194A Impedance Gain Phase Analyzer. The measurements were performed over a temperature range of 24 °C < T < 650 °C, during heating and cooling process, at a constant rate of 2 °C/min. The samples were cut with different geometries (bars and discs) to enable the piezoelectric characterization through the Gain-Bandwidth method.[11] Measurements of electrical resistivity vs temperature allowed to choose the optimal poling temperature (prior the slope of $log \rho$ vs T^{-1} increasing), which was 150 °C. All samples were poled in this temperature with an electric field of 4 kV/mm for 30 min.

RESULTS AND DISCUSSION

As mentioned in the introductory section, the fabrication of a PN ceramic with a single phase is a difficult task. Lee and co-authors[5,6] studied the phase evolution of the PN ceramics, obtaining the ferroelectric phase through a thermal treatment at 1300 °C and an appropriated quenching procedure before the sintering of the ceramic bodies. In this work, the orthorhombic phase was obtained without the quenching. The addition of PbO in excess, during the synthesis process, allowed us to obtain PN ceramics with a single orthorhombic phase, as illustrated in Figure 1.

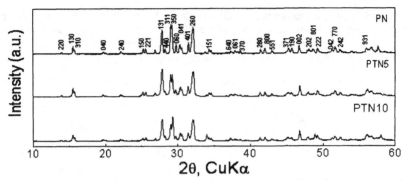

Figure 1. Room temperature X ray diffraction patterns of the PN and PTN powders, heat treated at 1300 °C during 1 h, showing only the orthorhombic ferroelectric phase.

Figure 2 shows the variation of the Curie temperature and the relative density of the sintered samples as a function of the Ti content. The density of PTN ceramics increased with the increasing of the Ti content. The Curie temperature also increased with the Ti content, from approximately 520 °C (for x = 0) up to 560 °C (for x = 0.1). On the other hand, the addition of Ba (x = 0.25) allowed the fabrication of higher densified ceramics when compared with the PTN, but decreased considerably the Curie temperature (~360 °C).

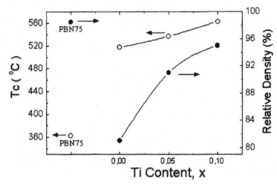

Figure 2. Curie temperature (T_c) and relative density, as a function of the Ti content, for the PTN ceramics. PBN75 data were added for comparison.

Figure 3 illustrates the room temperature mechanical quality factor (Q_m), for the thickness vibration mode of a thin disc and the relative dielectric permittivity ($\varepsilon_{33}/\varepsilon_0$), as a function of the Ti content, for the PTN and PBN75 ceramics. It can be observed that the room temperature permittivity decreased as the Ti content was increased. This characteristic is particularly desirable for high frequencies applications. It can be also observed, that the permittivity remained nearly constant, from room temperature up to approximately 400 °C, which provide a high stability when used in transducers. Otherwise, the addition of Ba increases the permittivity when compared to those doped with Ti. The permittivity at room temperature for PBN75 is about four times higher than that for PTN10. On the other hand, the quality factor is low for all samples (< 20), comparable with those of soft piezo-ceramics. Materials with a low quality factor are suitable for wide band ultrasonic transducers.

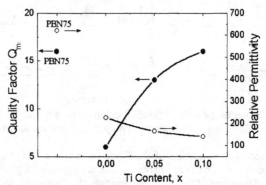

Figure 3. Quality factor (Q_m) and relative permittivity, at room temperature, as a function of the Ti content, for the PTN ceramics. PBN75 data were added for comparison.

Table I summarizes the main dielectric and piezoelectric properties for the studied ceramics. As can be observed, the longitudinal coupling factor (k_{33}) remains invariant, around 0.35, independently of the Ti or Ba content. The transverse coupling factor (k_{31}) could not be measured in the pure PN sample, because the piezoelectric activity was very small, inhibiting the necessary signals to the calculation. Nevertheless, this coefficient showed a progressive increasing when the Ti content was increased, diminishing the k_{33}/k_{31} ratio. On the other hand, the maximal permittivity (at the Curie temperature), decreased with increasing Ti content, contrary to the Ba doping, that favored the increase of this property. A large piezoelectric anisotropy (k_{33}/k_{31}) is desirable to improve the performance ultrasonic transducers that work in thickness mode, because of the practically absence of the planar vibrations influence. As can be observed in Table I, although the addition of Ti or Ba in PN ceramics decreased the piezoelectric anisotropy, it remained considerably higher than that observed in the commercial piezo-ceramics. Also, the relative permittivity, at room temperature, was lower than those reported, in reference 6, for pure PN ceramics.

Table I. Dielectric and electromechanical properties for $Pb_{1+x}Ti_xNb_{2-x}O_6$ (PTN100x) and $Pb_{0.75}Ba_{0.25}Nb_2O_6$ ceramics. k_{31}: transverse coupling factor, k_{33}: longitudinal coupling factor, Q_m: mechanical quality factor, for the thickness vibration mode, of a thin disc, T_c: Curie temperature, ε_{rt}: relative permittivity at room temperature, and ε_{max}: relative permittivity at the Curie temperature.

	Relative Density (%)	k_{31}	k_{33}	k_{33}/k_{31}	Q_m	T_c (°C)	ε_{rt}	ε_{max}
PN	81.0	-	0.35	-	6	517	227	6960
PTN5	91.2	0.016	0.36	22.5	13	537	165	6133
PTN10	94.9	0.031	0.35	11.3	16	563	140	5621
PBN75	98.5	0.063	0.34	4.9	16	363	622	7584
Ref 6*	~93	$0.049^{\ddagger}$	$0.35^{\pounds}$	$7.1^{\dagger}$	10	534	340	4000
Ref 6+	~96	$0.037^{\ddagger}$	$0.23^{\pounds}$	$6.2^{\dagger}$	32	540	430	4700

* Sample PN HT (C), sintered at 1250 °C.
+ Sample PN HT (F), sintered at 1250 °C.
‡ k_p
£ k_t
† k_t/k_p

CONCLUSIONS

High densities PN ceramics were fabricated using Ti and Ba, as doping elements. An alternative method was followed to obtain the room temperature ferroelectric orthorhombic phase. The addition of PbO excess, before calcination and sintering, compensated the lead mass losses during thermal treatments. In addition, an appropriate heat treatment at 1300 °C, during 1 h, allowed the obtaining of the orthorhombic phase, without the need of a quenching stage. Also, it was observed that the addition of Ti and Ba enhanced the densification of the PN ceramics. Ceramics of the composition with x = 0.1 reached 95% of the theoretical density, That is unusual by using conventional method. Ti doping also increased the Curie temperature and the quality factor and decreased the relative permittivity. Our results showed that PTN ceramics present high potential to be used in piezoelectric transducers to operate in high temperature and high frequencies. The improved piezoelectric anisotropy of the PTN ceramics is an additional feature for their potential for ultrasonic applications.

ACKNOWLEDGEMENT

The authors thank to Francisco J. Picon and CNPq/FAPESP brazilian agencies for the technical and financial support, respectively.

REFERENCES

[1] G. Goodman, "Ferroelectric properties of Lead Metaniobate", *J. Am. Ceram. Soc.*, **36** (11), 368 – 372 (1953).

[2] M. H. Francombe, "Polymorphism in Lead Metaniobate", *Acta Cryst.*, **9**, 683 – 684 (1956).

[3] E. C. Subbarao, "X Ray study of phase transition in ferroelectric $PbNb_2O_6$ and related materials", *J. Am. Ceram. Soc.*, **43** (9), 439 - 442 (1960).

[4]J. Soejima, K. Sato and K. Nagata, "Preparation and characteristics of ultrasonics Transducers for high temperatures using $PbNb_2O_6$", *Jpn. J. Appl. Phys.*, **39** (5B), 3083 – 3085 (2000).

[5]H. S. Lee and T. Kimura, "Sintering behavior of Lead Metaniobate" *Ferroelectrics*, **196**, 137 – 40 (1997).

[6]H. S. Lee and T. Kimura, "Effects of microstructure on the dielectric and piezoelectric properties of Lead Metaniobate", *J. Am. Ceram. Soc.*, **81** (12), 3228 – 3236 (1998).

[7]M. H. Francombe, "The relations between structure and ferroelectricity in lead barium and barium strontium niobates", *Acta Cryst.*, **13**, 131 – 140 (1960).

[8]C. V. Carmo, R. N. Paula, J. M. Póvoa, D. Garcia and J. A. Eiras, "Phase evolution and densification behavior of PBN ceramics", *J. Eur. Ceram. Soc.*, **19**, 1057-1060 (1999).

[9]T. Kimura, Y. Kuroda and H. S. Lee, "Effect of phase transformation on the sintering of lead barium metaniobate solid solutions", *J. Am. Ceram. Soc.*, **79** (3), 609 – 612 (1996).

[10]K. Nagata, Y. Kawatani and K. Okazaki, "Anisotropies of Hot-Pressed transparent (Pb, Ba, La) Nb_2O_6 ceramics", *Jpn. J. Appl. Phys.*, **22** (9), 1353 – 1356 (1983).

[11]R. Holland and E. P. EerNisse, "Accurate measurements of coefficients in a ferroelectric ceramic", *IEEE Transactions on Sonics and Ultrasonics*, **16** (4), 173 – 181 (1969).

ELECTRICAL PROPERTIES OF QUATERNARY PYROCHLORE RUTHENATES FOR THICK-FILM RESISTORS

K. Yokoyama, K. Kakimoto, and H. Ohsato
Materials Science and Engineering, Nagoya Institute of Technology
Gokiso-cho, Showa-ku, Nagoya 466-8555, Japan

J. Kinoshita and Y. Maeda
KOA Corporation
14016, Naka-minowa, Minowa-machi, Kamiina-gun, Nagano 399-4697, Japan

ABSTRACT

$Bi_{2-x}Ln_xRu_2O_7$ (Ln = Sm, Dy) have been synthesized and characterized by using X-ray diffraction, electrical resistivity measurements, and thermal expansion coefficient measurements. The lack of bismuth and oxygen site occupancy was observed in the composition $Bi_2Ru_2O_7$ by using Rietveld refinement. In the lanthanoid-rich, a change from metallic to semi-conducting behavior was observed at $0.8 < x < 1.0$ and $0.6 < x < 0.8$ in $Bi_{2-x}Sm_xRu_2O_7$ and $Bi_{2-x}Dy_xRu_2O_7$, respectively. The good average linear thermal expansion coefficient (α_l) 1.2×10^{-5} /K in the temperature range between 300 and 1100K was obtained by substituting lanthanoid, in the composition $Bi_{1.8}Sm_{0.2}Ru_2O_7$ and $Bi_{1.8}Dy_{0.2}Ru_2O_7$. It was superior to the value $\alpha_l = 3.2 \times 10^{-5}$ /K of $Bi_2Ru_2O_7$.

INTRODUCTION

Conductive materials with low thermal expansion of coefficient for avoiding stress and positive thermal coefficient of resistivity have been required for the development of thick film technologies as resistors. Most commercial, thick film resistors are based either on RuO_2 with the rutile structure, or $Bi_2Ru_2O_7$ or $Pb_2Ru_2O_{6.5}$ with the pyrochlore structure[1,2].

$Bi_2Ru_2O_7$ is characterized by high electronic conductivity and temperature-independent Pauli paramagnetism, while $Ln_2Ru_2O_7$ (Ln = Pr-Lu) and $Y_2Ru_2O_7$ are semiconducting with low activation energies[3,4,5]. The solid solutions, $Bi_{2-x}Ln_xRu_2O_7$ (Ln = Y, Pr-Lu) systems show metallic to semi-conducting property with composition[6]. Their structures have been refined by X-ray Rietveld analyses[7]. However, their electrical resistivities above room temperature and thermal expansion coefficient are not clear yet. In this study, the electric and thermophysic properties of $Bi_{2-x}Ln_xRu_2O_7$ (Ln = Sm, Dy) are evaluated. In addition, the crystal structure of $Bi_2Ru_2O_7$ was investigated by using Rietveld analysis.

EXPERIMENTAL

Samples were prepared by a solid-state reaction using Bi_2O_3, RuO_2 and Ln_2O_3 to form the pyrochlore-type $Bi_{2-x}Ln_xRu_2O_7$ solid solutions. The mixtures of the starting materials were ground, pressed into pellets and sintering in air at 1273-1423 K for 48h. The X-ray powder diffraction data were obtained using $CuK\alpha$ radiation (Philips, X'pert system FW3040). The lattice parameters were refined by whole-powder-pattern-decomposition (WPPD) method using the program WPPF[9]. Furthermore, $Bi_2Ru_2O_7$ structural parameters were refined by Rietveld analysis using the program RIETAN-2000[9]. The electrical resistivities were measured by a dc four-probe method between 300 and 400 K. The thermal coefficient of resistivity was obtained from the resistivity compare ρ_{300} to ρ_{400}. The thermal expansion of coefficient was calculated from the high-temperature powder X-ray diffraction analysis. This analysis was performed from 300K to 1100K.

RESULTS AND DISCUSSIONS

Electrical and thermophysical properties

Figure 1 shows the resistivities for $Bi_{2-x}Sm_xRu_2O_7$ and $Bi_{2-x}Dy_xRu_2O_7$ as a function of temperature. In the lanthanoid-rich, a change from metallic to semiconducting behavior is observed at $0.8 < x < 1.0$ and $0.6 < x < 0.8$ in $Bi_{2-x}Sm_xRu_2O_7$ and $Bi_{2-x}Dy_xRu_2O_7$, respectively.

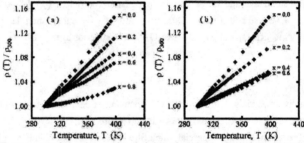

Fig. 1 Temperature dependence of resistivities for (a) $Bi_{2-x}Sm_xRu_2O_7$, (b) $Bi_{2-x}Dy_xRu_2O_7$.

The thermal coefficients of resistivity (TCR) are shown in Table 1. With increasing amount of lanthanoid, the TCR is decreased. According to the electronic structure calculate by W. Y. Hsu et al.[10], the conductivity of $Bi_2Ru_2O_7$ is caused by the unoccupied Bi 6p state and mixing with the Ru 4d state via the framework oxygen. As decreasing of amount for bismuth, influence of Bi 6p state is smaller and smaller. Therefore, it is found that the conductivities of $Bi_{2-x}Ln_xRu_2O_7$ decrease.

Table 1 Composition dependence of the thermal coefficients of resistivity for $Bi_{2-x}Ln_xRu_2O_7$.

Composition x	T.C.R. ($\times 10^{-3}$ /K)	
	Ln = Sm	Ln = Dy
0	1.5	1.5
0.2	1.1	0.97
0.4	0.88	0.57
0.6	0.71	0.54
0.8	0.29	negative
1.0-2.0	negative	negative

From the high-temperature X-ray analysis, the temperature dependence of lattice volume ($Å^3$) for $Bi_2Ru_2O_7$ is shown in Figure 2.

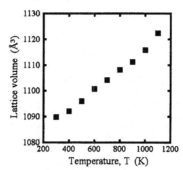

Fig. 2 Temperature dependence of lattice volume for $Bi_2Ru_2O_7$.

The lattice volume increases in a linear fashion with increasing temperature. The temperature dependences of linear thermal expansion coefficient $\alpha_l(K^{-1})$ is given from lattice volume as follows:

$$\alpha_l = -7.88 \times 10^{-15}T^3 + 7.50 \times 10^{-11}T^2 - 6.82 \times 10^{-8}T + 4.29 \times 10^{-5} \quad (1)$$

where T is the temperature[K]. The average α_l of $Bi_2Ru_2O_7$ is 3.2×10^{-5} /K in the temperature range between 300 and 1100K. From the same measurement, the average α_l of 1.2×10^{-5} /K is obtained for the both composition of $Bi_{1.8}Sm_{0.2}Ru_2O_7$ and $Bi_{1.8}Dy_{0.2}Ru_2O_7$. The average α_l of $Bi_{1.8}Sm_{0.2}Ru_2O_7$ and $Bi_{1.8}Dy_{0.2}Ru_2O_7$ are lower than that of $Bi_2Ru_2O_7$. Thus, slightly doped lanthanoids in $Bi_2Ru_2O_7$ have effects for decreasing the thermal expansion of coefficient.

Crystallographic

According to the XRD results, the patterns show that the solid solutions with the pyrochlore type single phase are synthesized. Figure 3 shows composition dependence of the lattice parameters of $Bi_{2-x}Ln_xRu_2O_7$, which is refined by using WPPD method. The pyrochlore structure $A_2B_2O_7$ is known that A-site is 8-coordination and B-site is 6-coordination. The ionic radius of Sm^{3+} (r=1.079Å: 8-coordination) and Dy^{3+} (r=1.027Å: 8-coordination) is smaller than that of Bi^{3+} (r=1.17Å: 8-coordination)[11]; therefore, if lanthanoid ions substitute for only bismuth site, the lattice parameters must be decrease as increasing lanthanoid ions. However, it did not followed the Vegard's Laws in the case of the $Bi_{2-x}Sm_xRu_2O_7$. The ionic radius of Sm^{3+} (r=0.958Å: 6-coordination) is larther than that of Ru^{4+} (r=0.620Å: 6-coordination); therefore, it is considered that a part of Sm ion is substituted for B sites with lack of oxigen. Differences of a change from metallic to semiconducting behavior range may be caused by modification of solution site.

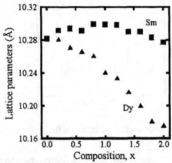

Fig. 3 Composition dependence of lattice parameters for $Bi_{2-x}Ln_xRu_2O_7$ (Ln = Sm, Dy).

To verify the occupancy of each site for the $Bi_2Ru_2O_7$, the Rietveld refinement carried out. Table 2 and Figure 4 shows the result of Rietveld refinement structure, which used the space group Fd-3m. The refined lattice parameter is similar to the previous reports[4,7,12]. The results are indicate that the lack of oxygen and Bi are observed as shown in Table 2, which are similar to the report of Carbonio et al.[13]. The linear thermal expansion coefficient of $Bi_2Ru_2O_7$ is too large for ceramics. This characteristic may be caused by the lack of bismuth and its high isotropic atomic displacement parameter.

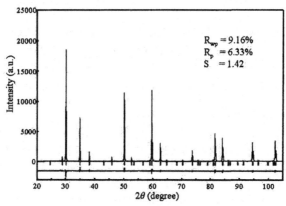

$R_{wp} = 9.16\%$
$R_p = 6.33\%$
$S = 1.42$

Fig. 4 Rietveld analysis of $Bi_2Ru_2O_7$.

Table 2 Crystalgraphic parameters of $Bi_2Ru_2O_7$.

Atom	Site	g	x	y	z	U ($Å^2$)
Bi	16d	0.934(2)	0.5	0.5	0.5	0.0176(2)
Ru	16c	1.0	0	0	0	0.0044(2)
O1	48f	1.0	0.3252(5)	0.125	0.125	0.015(3)
O2	8b	0.87(2)	0.375	0.375	0.375	0.015(3)

Space group : $Fd\bar{3}m$ a = 10.2902(2) (Å)

CONCLUSION

The $Bi_{2-x}Ln_xRu_2O_7$ solid solutions with the pyrochlore structure were prepared, and the electric and thermophysical properties were evaluated. A change from metallic to semi-conducting behavior was observed at $0.8 < x < 1.0$ and $0.6 < x < 0.8$ in $Bi_{2-x}Sm_xRu_2O_7$ and $Bi_{2-x}Dy_xRu_2O_7$, respectively. The linear thermal expansion coefficient was improved with slightly doped lanthanoids in $Bi_2Ru_2O_7$. From the Rietveld results, the lack of oxygen and bismuth were verified in the $Bi_2Ru_2O_7$.

ACKNOWLEDGEMENTS

This work was supported by a grant from the NITECH 21st Century of Excellence (COE) Program "World Ceramics Center for Environment Harmony" by the Japanese Ministry of Education, Science and culture.

207

REFERENCES

[1] G. E. Pike and C. H. Seager, "Electrical properties and conduction mechanisms of Ru-based thick-film (cermet) resistors," *J. Appl. Phys.*, **48**, 5152-5169 (1977)

[2] P. F. Carcia, A. Ferreti, and A. Suna, "Particle size effects in thick film resistors," *J. Appl. Phys.*, **53**, 5282-5288 (1982)

[3] R. J. Bouchard and J. L. Gillson, "A new family of bismuth - Precious metal pyrochlores ," *Mater. Res. Bull.*, **6**, 669-679 (1971)

[4] M. K. Hass, R. J. Cave, M. Avdeev, and J. D. Jorgensen, "Robust paramagnetism in $Bi_{2-x}M_xRu_2O_7$ (M = Mn, Fe, Co, Ni, Cu) pyrochlore," *Phys. Rev. B*, **66**, 99429 (2002)

[5] R. Aleonard, E. F. Bertaut, M. C. Montmory, and R. Paushenet, "Rare-Earth Ruthenates," *J. Appl. Phys.*, **33**, 1205 (1962)

[6] A. Ehmann and S. Kemmler-seck, "Systeme $Bi_{2-x}Ln_xRu_2O_7$ (Ln = Sm-Lu, Y, In)," *Mater. Res. Bull.*, **20**, 437-442 (1985)

[7] T. Yamamoto, R. Kanno, Y. Takeda, O. Yamamoto, Y. kawamoto, and M. Takano, "Crystal Structure and Metal-Semiconductor Transition of the $Bi_{2-x}Ln_xRu_2O_7$ Pyrochlores (Ln=Pr-Lu)," *J. Solod State Chem.*, **109**, 372-383 (1994)

[8] H. Toraya, "Whole-Powder Pattern Fitting Without Reference to a Structural Model: Application to X-ray Powder Diffractometer Data," *J. Appl. Cryst.*, **19**, 440-447 (1986)

[9] F. Izumi and T. Ikeda, "A Rietveld-analysis program RIETAN-98 and its applications to zeolites," *Mater. Sci. Forum*, **198**, 321-324 (2000).

[10] W. Y. Hsu, R. V. Kasowski, T. Miller, and T. C. Chang, "Band structure of metallic pyrochlore ruthenates $Bi_2Ru_2O_7$ and $Pb_2Ru_2O_{6.5}$," *Appl. Phys. Lett.*, **52**, 792-794 (1988)

[11] R. D. Shannon "Revised effective ionic radii and systematic studies of interatomic distances in halides and chalcogenides," *Acta Cryst.*, A32, 751 (1976)

[12] G. R. Facer, M. M. Elcombe, and B. J. Kennedy, "Bismuth Ruthenium Oxides. Neutron Diffraction and Photoelectron Spectroscopic Study of $Bi_2Ru_2O_7$ and $Bi_3Ru_3O_{11}$," *Aust. J. Chem.*, **46**, 1897-1907 (1993)

[13] R. E. Carbonio, J. A. Alonso, and J. L. Martinez, "Oxygen vacancy control in the defect $Bi_2Ru_2O_{7-y}$ pyrochlores: a way to turn the electronic bandwidth," *J. Phys.: Condens. Matter*, **11**, 361-369 (1999)

MEASUREMENT OF COMPLEX PERMITTIVITY OF LOW TEMPERATURE CO-FIRED CERAMIC AT CRYOGENIC TEMPERATURES

Mohan V. Jacob
Electrical And Computer Engineering, James Cook University, Townsville, Australia

Janina Mazierska
Institute of Information Sciences, Massey University, Palmerston North, New Zealand
and
Electrical And Computer Engineering, James Cook University, Townsville, Australia

Marek Bialkowski
School of Information Technology and Electrical Engineering, University of Queensland, Australia

ABSTRACT

Precise knowledge of microwave properties of Low Temperature Co-fired Ceramic *LTCC* materials is crucial for efficient design of microwave systems, especially for design of communication filters. Many communication devices operate at low temperatures. In order to implement LTCC at low temperature communication circuits and devices, Engineers need precise values of the complex permittivity. The aim of this paper is to characterise LTCC material as a function of temperature at microwave frequencies using a very precise measurement technique, Split Post Dielectric Resonator. The Transmission Mode Q-factor technique that eliminates all the parasitic losses during the characterisation was used for the S-parameter data processing. The permittivity and loss tangent data of LTCC manufactured by Heraeus Circuit Materials Division in the temperature range 20 K – 300K at a frequency of 9.5 GHz is reported in this paper. The random uncertainty in ε_r and $\tan\delta$ are better than 0.3% and 5 % respectively.

Keywords: Dielectric Properties; LTCC; Substrates

INTRODUCTION

The most desired factors like pocket size, light weight, low power, low cost have triggered a revolution in the development of wireless communication systems which demand better performance than earlier systems in a multipath environment[1]. In modern electronic systems passive components are several fold higher than active components. Example in GPS the ratio of passive to active components is higher than hundred[1]. Low Temperature Co-fired Ceramic (LTCC) technology allows integration of these passive components in electronic systems and as a result the size reduction. All layers can be now processed in parallel, reducing the production cost and time. Lower temperature firing of ceramic blocks allows utilization of highly conductive metals such as gold or silver and decrease of line-loss[1-4]. As a result of this progress a very rapid growth of applications of LTCC in wireless communications has been observed during the last few years.

The Co-fired Ceramic multilayer technology allows for fabricating three-dimensional circuits, in which resistors, inductors, capacitors and low loss interconnects are "buried" within a ceramic block[1,4]. The LTCC technology (low firing temperature of ceramic blocks) enables utilisation of low loss metals such as gold or silver, making very attractive technology for RF and microwave applications. The main advantages of LTCC technology are: parallel processing, precisely defined parameters and stable performance over the lifetime, high performance conductors, and very high density of interconnects[2,4]. Compared to conventional PCB materials, LTCC offers a significantly higher thermal conductivity[5]. Due to these advantageous features, the LTCC technology is very attractive for developing wireless communication circuits such as transmit/ receive switches, delay lines, filters, impedance matching VCO and directional couplers. LTCC allows integration and miniaturization of passive circuits in electronic circuits containing active elements. Low Temperature Co Fired Ceramic using the stacked LTCC technology, compact internal multi-band chip antenna can be achieved for mobile communication handsets[6].

Table I Available LTCC in the market[1]

Company	Identification	Permittivity
Dupont	951	7.5
	934	7.8
Electro-Science	41110-25C	4-5
Laboratories	41010-25C	7.2-8.2
	41020-25C	8-10
	41110-70C	4.3-4.7
	41020-70C	7-8
Ferro	A6M	5.9
	A6S	5.9
Heraeus	CT2000	9.1
	Ct700	7.5-7.9
	CT800	7.5-7.9
Kyocera	GL550	5.6-5.7
	GL660	9.4-9.5
Nikko	Ag2	7.8
	Ag3	7.1
Northrop Grumman	Low K	3.9
Samsung	TCL-6A	6.3
	TCL-7A	6.8

Table I shows the commercially available LTCC in the market. Low temperature complex permittivity data of LTCC is not available except for a limited range of temperatures. We have developed a Split Post Dielectric Resonator, which can be cool down to temperature of 20 K. In this paper we have characterized four LTCC materials of relative permittivity between 6.8 and 9.2 as a function of temperatures (from 25 K to 290 K) at 9.5 GHz frequency.

EXPERIMENTS
Higher accuracy in microwave chracterisation of substrate materials can be attained by using Split Post Dielectric Resonator (SPDR) technique[8-12]. The substrate under test is placed between two low

loss dielectric rods situated in a metallic enclosure, as shown in Fig. 1. Eventhough different modes of the resonator can be identified and used for the microwave characterisation, $TE_{01\delta}$ mode has been used since this mode is insensitive to the presence of air gaps perpendicular to z-axis of the resonator.

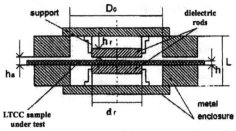

Fig. 1 Split-post resonator

The complex permittivity has been calculated based on the rigorous electromagnetic modeling of the split post resonant structure using the Rayleigh-Ritz technique[12]. A computer program has been developed for the calculation of complex permittivity. The real part of the sample's complex permittivity is computed from measured resonant frequencies of the resonator using the following equation[12]:

$$\varepsilon'_r = 1 + \frac{f_0 - f_s}{h f_0 K_\varepsilon (\varepsilon'_r, h)} \tag{1}$$

where: h is thickness of the sample under test, f_o is the resonant frequency of the empty SPDR, f_s is the resonant frequency of the resonator with the dielectric sample. K_ε is a function of ε_r' and h, and has been evaluated for a number of of ε_r' and h using Rayleigh-Ritz technique. Iterative procedure is then used to evaluate subsequent values of K_ε and ε_r' from equation (1).

The loss tangent of the tested substrate is calculated from the measured unloaded Q_0-factors of the SPDR with and without the sample based on:

$$\tan \delta = (Q_0^{-1} - Q_{DR}^{-1} - Q_c^{-1}) / \rho_{es} \tag{2}$$

where p_{es} electric energy filling factor of the sample, Q_{DR}^{-1} and Q_c^{-1} denote losses of the metallic and dielectric parts of the resonator respectively.

The measurement system used for the microwave characterisation of the LTCC materials sample is shown in Fig. 2. The system consisted of Network Analyser (HP 8722C), closed cycle refrigerator (APD DE-204), temperature controller (LTC-10), vacuum Dewar, a PC and the Split-Post dielectric resonator.

For variable temperature measurements S-parameter data sets were measured first for the empty resonator and then for the resonator with a given LTCC sample. To obtain precise values of the Q_0-factor of the split-post resonator and hence accurate values of tanδ of LTCC substrates we have measured many points around the resonance and processed measured data sets using the Transmission Mode Q-Factor Technique[13]. The TMQF technique was then used to obtain f_o and Q_o values of the empty split post resonator and of the resonator with the LTCC sample, at exactly the same temperatures. The microwave parameters ε_r and $\tan\delta$ were computed from the resonant frequencies and unloaded Q_o-factors respectively using a software developed.

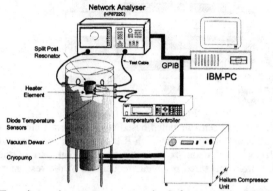

Fig. 2. Experimental set-up to measure the Q-factor and f_o of LTCC SPDR

RESULTS AND DISCUSSION

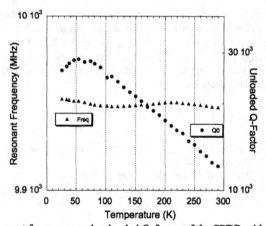

Fig. 3 The resonant frequency and unloaded Q-factor of the SPDR without sample.

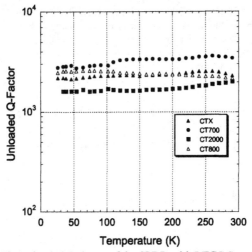

Fig. 4 The unloaded Q-factor of the SPDR with LTCC Samples

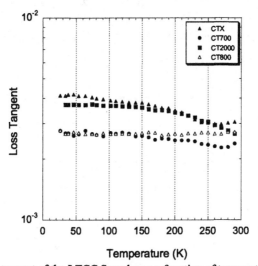

Fig. 5 The loss tangent of the LTCC Samples as a function of temperature at 10 GHz.

Four LTCC materials having different permittivity are characterised as a function of temperature (from 25 K to 290 K) at microwave frequencies (9 GHz). The thicknesses of the samples vary between 0.56 mm and 0.75 mm (CT700 – 0.720 mm, CT800 – 0.750mm, CT2000 – 0.700 and CTX – 0.556mm). Fig. 3 shows the resonant frequency and unloaded Q-factor of the empty resonator at different temperatures. The resonant frequency and Q_0-factor shows peaks at certain temperatures due to the presence of BMT dielectric in the cavity.

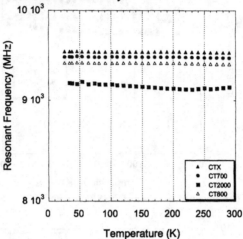

Fig. 6 The resonant frequency of the SPDR with LTCC Samples at different temperatures.

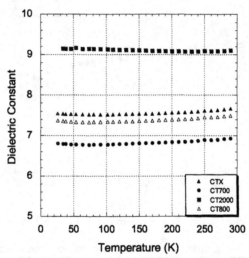

Fig. 7 The real part of the complex permittivity LTCC Samples at different temperatures.

Fig. 4 shows the unloaded Q-factor of the LTCC samples at different temperatures. The loss tangents of the samples were calculated using eq. (2) for each temperatures and represented in Fig. 5. CT700 and CT800 exhibit the smallest loss and the variation in losses due to the temperature is significantly small. This shows that the insertion loss of devices fabricated using these materials will be stable at varying temperatures. Fig. 6 shows the resonant frequency variation of the SPDR with LTCC samples. The real part of permittivity is estimated from the f_0 data using eq. (1) and is graphically represented in Fig. 7. The variation in permittivity is really small in the temperature range 25-290K. We have estimated the temperature coefficient of permittivity (τ_ε) from the calculated values of permittivity and is shown in Fig. 8. The τ_ε of sample CT2000 is really small. Therefore fluctuation in resonant frequency due to the temperature variation will be small for devices fabricated using CT2000 LTCC.

To access accuracy of our measurements we performed uncertainty analysis of measured ε_r and tanδ. Due to the uncertainty of the sample thickness we estimated uncertainty in real part of relative permittivity is around 0.3%. The random relative uncertainty in loss tangent was calculated to be at most 5% assuming 3% uncertainty in the Q_0-factor values.

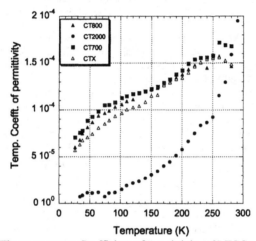

Fig. 8 The temperature Coefficient of permittivity of LTCC samples.

CONCLUSIONS

The real part of relative permittivity and loss tangent of four types of Low Temperature Co-Fired Ceramics is precisely characterised at 9.5 GHz in the temperature range 25K – 290K. The measurements were performed using a cryogenic split post dielectric resonator. The tested LTCC samples exhibited permittivity values of 9.1 (CT2000), 7.6 (CTX), 7.4 (CT800) and 6.9 (CT700) at room temperature. The corresponding loss tangents of the samples were 0.0026, 0.003, 0.0027 and 0.0024. The dependence of tanδ and permittivity were very small in the temperature range 25K to

290K. Therefore we anticipate stable operation of LTCC devices and circuits at varying temperatures.

ACKNOWLEDGEMENTS
This work was done under the financial support of ARC Discovery Project DP0449996. The authors acknowledge the samples from Heraeus Circuit Materials Division. MVJ also acknowledges the ARC Australian Research Fellowship.

REFERENCES
[1] J.F. Kiang, "Novel Technologies for millimetre wave applications" (refer pp. 173-190), Kluwer Academic Publications, USA (2003).

[2] H. Mandai, K. Wakino and N. Nakajima, Recent Development on materials for multiplayer ceramic microwave devices, APMC2001 SMMM, 2001, Taipei, Taiwan, 142-145 (2001).

[3] M. Valant, Glass-free materials for LTCCC technology, 2001, APMC2001 SMMM, 2001,Taipei, Taiwan, 6-11 (2001).

[4] P. Barnwell., C. Free, and Z. Tian, Low Temperature co-fired ceramics (LTCC) and thick film technologies for microwave applications to 70GHz, 2001, APMC2001 SMMM, 2001,Taipei, Taiwan, 1-4 (2001).

[5] U. Langman, P. Mayr, S. Mecking, "Wireless RF multi-chip modules for LTCC system in package solutions: Key design considerations", Proceedings of APMC 2004, New Delhi (2004).

[6] Y. Kim, H.Lee, "Design of dual band meander chip antenna with gap-stub for mobile handsets", Proceedings of APMC 2004, New Delhi (2004).

[7] J. Mazierska, M. V. Jacob, A. Harring, J. Krupka, P. Barnwell and T. Sims, "Measurements of loss tangent and relative permittivity of LTCC ceramics at varying temperatures and frequencies", European Journal of Ceramic Society, 23 2611-2615 (2003).

[8] J. DelaBalle, P. Guillon and Y. Garault, Local complex permittivity measurements of MIC substrates. *AEU Electronics and Communication,* 35, 80-83 (1981).

[9] T. Nishikawa *et al.,* Precise measurement method for complex permittivity of microwave substrate. *CPEM '88,* 154-155 (1988).

[10] J. Krupka, R. Geyer, J. Baker-Jarvis and J. Ceremuga, Measurements of the complex permittivity of microwave circuit board substrates using split dielectric resonator and reentrant cavity techniques. In *DMMA96 Conference,* Bath, UK, 21-24 (1996).

[11] J. Krupka, S. Gabelich, K. Derzakowski and B. M. Pierce, Comparison of split post dielectric resonator and ferrite disc resonator techniques for microwave permittivity measurements of polycrystalline yttrium iron garnet. *Meas. Sci. Technol.,* 10, 1004-1008 (1999).

[12] J. Krupka, A.P. Gregory, O. C. Rochard, R.N. Clarke, B. Riddle and J. Baker-Jarvis, "Uncertainty of complex permittivity measurements by split post dielectric resonator technique", Europ. J. Ceram. Soc., 21, 2673-2676 (2001).

[13] K Leong, J Mazierska, Precise Measurements of the Q-factor of Transmission Mode Dielectric Resonators: Accounting for Noise, Crosstalk, Coupling Loss and Reactance, and Uncalibrated Transmission Lines, *IEEE Trans. Microwave Theory and Techniques,* 50, 2115 –2127 (2002).

Author Index

9 781574 982350